The Animals Among Us

The Animals Among Us

The New Science of Anthrozoology

JOHN BRADSHAW

ALLEN LANE
an imprint of
PENGUIN BOOKS

ALLEN LANE

UK | USA | Canada | Ireland | Australia
India | New Zealand | South Africa

Penguin Books is part of the Penguin Random House group of companies
whose addresses can be found at global.penguinrandomhouse.com.

First published in the United States of America by Basic Books, a member of the
Hachette Book Group 2017
First published in Great Britain by Allen Lane 2017
001

Printed in Great Britain by Clays Ltd, St Ives plc

A CIP catalogue record for this book is available from the British Library

ISBN: 978–0–241–18463–9

www.greenpenguin.co.uk

Penguin Random House is committed to a
sustainable future for our business, our readers
and our planet. This book is made from Forest
Stewardship Council® certified paper.

To my lovely wife, Nicky,
who introduced me to the joys of keeping pets.

Contents

Preface

W HY KEEP PETS? Such a question will never have occurred to some people: there has always been a dog at home, or a cat, or maybe rabbits or guinea pigs. So many households in the West include pets nowadays that it's salutary to note that a mere two centuries ago, when domestic animals were everywhere, pet keeping was almost exclusively the prerogative of the rich. Of course, increasing affluence has allowed many more people to keep animals as companions, but that alone cannot explain the massive surge in pet ownership in the last century. Survey after survey shows that people feel deep emotional attachment to their pets and think of them as 'one of the family'.

Where does this passion come from? I believe that it stems from deep within our nature and provides insight into what makes us human. Many of the 'Stone Age' tribes whose way of life survived into the twentieth century kept animals as companions. A capacity to regard animals as friends and not merely as food may therefore be as ancient as our species. It has never become universal, however: lots of people feel no attachment to animals or inclination to keep them as pets. Any theory as to why pets are such an important part of many people's lives must take account of that fact.

I CAME TO WRITE this book as a scientist as much as an animal enthusiast. Indeed, I started life as something of a pet agnostic. My father didn't like cats or dogs, so there were no pets in my childhood,

apart from those I encountered when visiting relatives and neighbours. As a biologist in the making, I saw pets mainly as interesting
and accessible animals, not all that different from the birds that vis-
ited the feeders in our garden or the deer that we'd sometimes disturb
on family walks in the countryside where we lived (we'd inherited
the tradition of walking – with or without a dog – from my mother's family). My wife, who had grown up with dogs, cats, and, at one
point, an unintended greenhouse full of rabbits, introduced me to the
joys of greetings from a wet nose and wagging tail – not to mention
what still feels like the privilege of sharing a home with a purring but
undoubtedly independent felid.

As I became a professional biologist, I initially studied animals,
not people – and certainly not the relationship between the two.
Even when studying dog and cat behaviour, I found myself somewhat
professionally isolated. Back in the 1980s, the study of any kind of
domestic animal, let alone the ultimate artificialities that are pet
dogs and cats, was deeply unfashionable. A brief flurry of interest
in their biology in the 1960s, when vets like Dr Michael Fox established some of the basics – such as the need to expose domestic dogs
to the right kind of contact with people during their 'socialization
period' – had quickly died away. The current surge of interest in dog
behaviour would have been unimaginable in the early stages of my
career.

I was initially drawn into studying dogs, and then cats, because
I was researching the ways that animals of all kinds use scent to
communicate. Theirs is a world of smelly information to which we
humans, with perhaps the least sensitive noses of any mammal, are
oblivious. I soon became fascinated by the behaviour of both cat and
dog, and not just the part governed by scent. I came to recognize in
it an enthralling compromise between their wild origins as canid and
felid and the requirements of their modern domestic lifestyles. And I
became increasingly aware that their behaviour could not be divorced
from how their owners behaved and thought about them.

It soon dawned on me that, although I might think about cats
and dogs as mammals while at work, at home neither I nor anyone

else viewed our pets in this way. Most owners treat these animals as little people, talking to them as if they understand what is being said. I seemed to do this as much as anyone else, despite being well aware that such comprehension was almost certainly impossible.

Once I began to think about pets in their domestic context, it followed that although the various purposes we have put them to down the ages have undoubtedly shaped the dogs and cats of today, this cannot be the whole story. The subconscious workings of our brains have also moulded them, and now that we have largely liberated them from their former utilitarian roles, these influences are becoming paramount. Their cuteness attracts us (well, most of us) instinctively: the features that kittens and puppies, and to a lesser extent adult cats and some dogs, share with human babies seem to trigger our protective and caring side. Our brains also perceive intention in almost anything that moves, and thus without a moment's reflection we automatically ascribe all kinds of humanlike emotions and intentions to our animals. Some of these are probably reasonably accurate; others are wildly imaginative. Generally the errors don't matter, but sometimes they do, as when our pets fall short of the often unrealistic expectations we have of them. I eventually concluded that we simply can never understand the behaviour of pet animals without factoring in the workings of the human mind.

This realization kindled my interest in what the few academics with any interest in the subject back in the 1980s referred to as the 'human–animal bond'. I began to attend conferences that included the topic of pet ownership. At a workshop on the day after one such conference – as it happened, also my fortieth birthday – I and the six other academics present coined the term 'anthrozoology' to sum up what we all did and to give a distinctive title to the society we founded a year later, the International Society for Anthrozoology (not that we could take credit for creating this word from scratch: we had adapted it from the title of an academic journal, *Anthrozoös*, which had published its first issue three years previously). The word stuck, and now, a quarter of a century on, students can earn a qualification in anthrozoology from universities all around the world. The field mainly brings together various types of scientists – biologists, vets, medical researchers, psychologists, social scientists and anthropologists – who study the personal relationships that people have with animals and, to a lesser extent, that animals have with people. Thus anthrozoology has focused most on those types of animals that serve widely as companions: dogs, cats, horses, aquarium fish, riding horses, and so on. (A parallel movement, sometimes referred to as human–animal studies, has grown up among those interested in more philosophical approaches, such as ethics, political science, history, geography and literature. There is, unfortunately, not much dialogue between the two camps.)[1]

In some senses, I've been writing this book for more than twenty years. The more I studied pets and their owners, the more intrigued I became. Despite sometimes suffering due to our misunderstanding of their requirements, cats and dogs were obviously benefitting from their relationships with people – there are far more of them in the world today than there are of the wild ancestors from which they emerged thousands of years ago. Yet, somewhat paradoxically, their numbers actually grew just as their usefulness to us disappeared. Only a few dogs still herd sheep or help with hunting; most owners find their cat's once valued hunting prowess appalling. Pets' hold on their owners seems to run deeper than mere fashion. So, using as wide a range of sources as I could find, in addition to as much original research as I could fit in around my studies of cat and dog welfare, I set out to examine what science has to say about humans' fascination with animals and why we took them into our homes. This book is the result.

Conventions

IN THIS BOOK, I have resisted following the growing practice, in the United States especially, of referring to pets as 'companions', and throughout I refer to 'owners' rather than 'guardians' or 'parents', labels preferred by some authorities. Perhaps I'm old-fashioned, but to me, pet 'ownership' implies responsibility and certainly does not suggest a right to treat an animal like the inanimate sort of possession. I have a concern that the term 'guardian' implies a legal status that arises from some mental deficiency in the animal – hardly an appropriate way to characterize the relationship between pet and owner. 'Caregiver' – another suggestion I've seen – seems too impersonal and transient, more appropriate for those devoted souls who look after pets in rehoming facilities. 'Pet parent' is simply too anthropomorphic for me to stomach: for a biologist like myself, the term 'parent' indicates (though not exclusively) a genetic relationship between mother or father and offspring. In this book I also occasionally but unashamedly refer to a (singular) pet as 'it', but only when I don't know whether the animal I'm referring to is male or female: I mean no disrespect by this. I just find 'he or she' and 'him or her' unnecessarily clumsy.[1]

Many ethicists regard the term 'pet' as derogatory, preferring the more neutral 'companion animal'. True, the word 'pet' has other meanings: in the sixteenth century it referred to the subject of any favoured relationship, be it with an animal or another person, and even today in the northeast of England, 'pet' crops up in conversation as a (slightly condescending?) term of endearment directed at women and children – as in 'How're you doing today, pet?' However, it's not clear, at least to me, what advantage persuading the general public to refrain from using the word 'pet' (even if that were possible) would bring.[2]

Introduction

O NCE UPON A TIME there were three little pigs. One lived in Connecticut, one in Des Moines, Iowa and one in the Bismarck Mountains of New Guinea. Unlike those in the children's tale, all three were real. None ever built its own house or had any kind of altercation with a wolf. The humans around each thought it special, and fed and treated it well. We might regard all three, in their way, as pets, but each was also something else.

The pig from Connecticut attained brief notoriety the day before Thanksgiving 2014, when it (with its owner, aspiring model Rachel Boerner) was ejected from a US Airways flight before take-off for 'disruptive behaviour' (the pig had defecated in the aisle). The other passengers wondered why the airline had ever allowed a full-grown pot-bellied pig on a plane in the first place. A company representative clarified that the airline had mistakenly cleared the pig for take-off as an 'emotional support animal' (a somewhat nebulous concept since most pet owners would claim that their animals provide them with 'emotional support'). Whatever the ins and outs, this was a seventy-pound pig on a lead, not a toy dog in a handbag. Nonetheless, the animal's owner clearly regarded it as her inseparable companion.[1]

The pig from Des Moines – a $900 'therapy piglet' named Stuart – lived with a six-year-old boy rendered completely blind and partially deaf by a genetic disorder. The boy's mother, hoping that the company of an animal would comfort her son, first tried a dog but

1

An 'emotional support' pig on a plane.

found that its hair made her son gag. Stuart, however, was an instant success – as reported on *KCCI News*, the boy spontaneously wrapped his arms around the piglet on their first meeting, and thereafter they were firm friends.[2]

The New Guinea piggy served as both treasured companion and then ceremonial meal. Among the Maring people of the Bismarck Mountains, the family unit comprises both humans and pigs. Women care for both piglets and human infants, carrying porcine infants everywhere, sometimes swaddled in banana leaves to prevent them from struggling. The women train the growing piglets to follow them wherever they go. For their first nine months, the piglets live cheek-by-jowl with the household's children, after which they move into their own stalls within the family hut. During the day, while the women tend the gardens that provide the Maring with much of their food, the pigs scavenge in the forest, walking home to the village in the evening with the children. As their numbers increase, the pigs find it harder to obtain all the food they need by foraging, and so the women supplement their diet with sweet potatoes grown in the gardens.[3]

Eventually the number of pigs becomes a problem, and the women start to complain. Because the soil is poor, their gardens only produce for two or three years, after which the women must clear a new patch of forest. Each is inevitably further from the village than the last, so the women spend more and more time travelling and less time growing food. The hungry pigs may then roam as far as the neighbouring settlements and cause damage there.

At this point the men decide that the time is right to hold a *kaiko*, an extended festival that involves the sacrifice and consumption of most of the pigs. The villagers ritualistically club them to death, one by one, and then roast them, consuming vast amounts of pork over several weeks in ceremonies designed to re-establish family ties, build allegiances with neighbouring tribes, and arrange marriages for the unattached. After the feasting, the few remaining female pigs form the nucleus of the next generation, and so the cycle begins again, culminating some ten to fifteen years later in the next *kaiko*.[4]

These three pigs, equally well cared for during their lives and all companions in their way, enjoy three distinct kinds of relationships with humans. In the West, the pig that becomes a pet is highly unlikely to end up on the dinner table; yet in New Guinea this is the norm. Historically, people would have considered the idea that an animal could have therapeutic powers outrageous. And yet, in the past few decades, particularly in the West, many pet owners have come to believe just that.

However different the stories of these three pigs and their people, collectively they show that some aspect of human nature enables us to develop personal relationships with members of other species. True, the supermarket shelves demonstrate that we can treat animals as commodities when necessary, but most of us also relate to (other) animals as individuals. We surround ourselves with their pictures – the Internet, if not entirely made of cats, would certainly be a less salubrious place without them – and many of us cannot resist touching cats and dogs, even if they're unknown to us. We seem to have a profound need to reconnect with the hands-on involvement with animals that was a part of the everyday lives of our hunter-gatherer ancestors.

Today we relate to animals very differently than our forebears did. Our thinking about animals has changed dramatically over the past century or so. Nowadays, we accord many kinds of animals 'rights'. Companion animals – especially dogs and cats – are more popular than ever. Over the past forty years, the idea has emerged, seemingly from nowhere, that having an animal in the home is part of a healthy lifestyle. We view pets as the antidote to the stresses of modern urban living, as a palliative for the loneliness that comes with the demise of the extended family, as 'co-therapists' for all manner of disabilities. I find the speed and enthusiasm with which we have embraced these ideas – to the point where few question them, even though the science underpinning them is equivocal at best – more fascinating than the ideas themselves. Something in human nature seemingly encourages us to accept these assertions, as if pets deserve immunity from the scepticism that usually greets claims of hitherto untapped healing powers. We don't simply feel the urge to keep pets; we seem to want to believe that they provide some kind of elixir that will allow us to lead more fulfilling lives.

OVER THE PAST half-century or so, our relationships with animals have become more complex and thus more difficult to disentangle. With the exception of a small number of animals kept by aristocrats solely for the purpose of companionship, domestic animals generally filled a practical role first and then, occasionally, an emotional one as well. Horses transported; cats moused. Dogs have served many functions, including hunting, herding and guarding, thereby earning more opportunities to form bonds with individual people than any other species. A few species provide only a modicum of companionship or at least aesthetic pleasure: aquarium fish have remained popular, but cage birds – once among the most popular household pets and now the province of specialists – have all but disappeared. (One theory holds that with the invention of the radio, the housebound no longer needed them to fill the silence. Others point more cynically to the parallel increase in the popularity of indoor cats.)[5]

Some animals continue to straddle the line between utility and pure companionship. In many rural areas, animals kept primarily as pets often have some practical use – consider the weekend gun dog. And some that are primarily useful – for example, farm cats – are often objects of considerable affection. I once knew a retired shepherd whose relationship with his dogs was, on the surface, utilitarian in the extreme – he never even named his succession of collies, referring to each as 'my dog' (he communicated with them in the traditional way, purely with whistles). However he was clearly devoted to his final dog, which he had kept after hanging up his crook for the last time. He walked her past my house three times every day, in rain, hail and snow. Bert, on whose Oxfordshire farm I made my first studies of farm cat colonies, clearly had a soft spot for his eighty-odd felines, although he used but a single name – 'Ginge' – for all of them.

In general, however, the line between companion and domestic animals has become more sharply defined in recent decades. Nowadays in the West we distinguish sharply between those animals which we keep for our consumption and those that are purely our friends. Whereas the Maring eventually eat their 'pet' pigs, the pig owner prevented from flying home on Thanksgiving would most likely recoil from the thought of her pet as a source of bacon.

Traditionally, however abhorrent Western pet owners today may find it, both cats and dogs were bred not for their company but for their fur and meat. For dogs at least, such uses go back a long way. In central Europe in around 6000 BCE, the custom of burying dogs alongside their masters had been established for hundreds of years, but other communities nearby were routinely killing and eating dogs, as evidenced by broken long bones and skulls. During the Middle Ages, the English prized cat-skin coats and, as evidenced by cut marks on cat bones recovered from rubbish pits, deliberately culled cats for this purpose as soon as they were fully grown, at about one year old.

As recently as fifteen years ago, the Royal Society for the Prevention of Cruelty to Animals estimated that Chinese producers were sending as many as 2 million cat and dog pelts annually to European (and American) fur markets, although many of them probably ended up in Russia, where consumers are less fastidious. South Korea has a long dog-eating tradition and kills over a million canines each year for food; one can go into a popular chain of dog-meat restaurants, buy dog meat 'sweets', or even purchase a range of cosmetics derived from dog by-products. Koreans have less affinity for cats but do make a tonic for rheumatism and neuralgia by rendering a cat in a pressure cooker and blending the resulting broth with herbs. The inclusion of cats and dogs in their diet remains popular with many South Koreans, who may perceive strident calls from the West to give up this 'barbaric' practice as an attack on their cultural identity.[6]

In Western society today, pets do not get eaten. When someone becomes fond of an animal destined for the dinner table, translocating it into the 'pet' category, the consequences can be considerable. In the United Kingdom, during World War II and the years of meat rationing that followed, the government encouraged households to keep rabbits for the pot. In the words of one wartime child, written some seventy years later, 'I grew fond of one rabbit, a beautiful white Angora with pink ears and eyes. But like all the others, once it was fat enough, it was killed and strung up to eat. I have never been able to eat meat since those days and have been vegetarian ever since.'[7] Even in parts of the world where dogs and cats are eaten, owners do seem to distinguish between those raised as pets and those bred for the pot.

AT A TIME in history when our need for cats and dogs as working animals has declined sharply, our urge to keep them is actually growing. Dogs and then cats are far and away the most popular household pets. Almost a quarter of households in the United Kingdom and over a third in the United States have one or more dogs, and cats share a roof with 30 per cent of US families and about 17 per cent of UK families. Rabbits, cage birds, guinea pigs, hamsters and assorted reptiles make up most of the other animals kept in British and American homes. Many horses, poultry and other livestock kept outdoors may also be regarded as companions. In Japan, insects such as stag beetles, crickets and fireflies, collectively known as *mushi*, are popular with children, especially boys. If we include fish, the United Kingdom has roughly as many animal companions as humans; the United States has about three for every four people.[8]

Pet keeping has metamorphosed from a pastime of the leisured classes into an integral part of the lifestyle of people from all social strata. Consumerism makes the practice easier, because products specifically designed for a pet's needs are now readily available 24/7. Advertisements for pet food and other products, while targeting pet owners, inevitably feature desirable animals that must also boost the overall image of pet keeping.

With globalization, Western-style pet keeping has spread into the Far East, where it is rapidly replacing local traditions. In Korea, dogs and cats (mainly Western breeds) have become popular pets since the 1980s, spawning a pet-products industry that culminated in the 2003 opening of Mega Pet, an eleven-story pet department store near Seoul. In China, while some dogs are still bred for their meat, imports from the West, such as poodles and many toy breeds, make popular pets for city dwellers: the days when Chairman Mao denounced pet keeping as a bourgeois indulgence seem to have passed, perhaps forever. In big cities like Beijing and Shanghai, pet dogs have become so popular that the law restricts owners to one per household and bans large breeds. There are rumours that some owners keep large dogs exclusively indoors – almost certainly compromising their welfare – simply to avoid their detection by police.[9]

The money that owners spend on their pets has reached staggering proportions. Whereas dogs and cats once lived on household scraps, nowadays a multi-million-dollar industry has grown up to service their needs. In 2014 UK owners forked out around £6 billion on their pets, much of this on food but over a third on veterinary services that barely existed half a century ago, when veterinary surgeons primarily worked with livestock. In the same year, US owners spent an estimated $60 billion, up 5 per cent from the previous year.

The latest pet-food trends track those in human food – gluten-free, low-sugar, reduced-calorie – reflecting not simply the genuine issue that many of today's pets are obese but also their owners' belief that their pets should eat as well as they do. Meanwhile, as campaigners against global hunger point out, a similar amount of money, if distributed fairly, could potentially feed every malnourished person in the world.[10]

BECAUSE OUR RELATIONSHIP to our domestic animals has changed over the past couple of hundred years, the term 'pet' may require some clarification. *Webster's Dictionary* defines a pet as 'a domesticated animal kept for pleasure rather than utility'. For my purposes, this distinction is somewhat restricting since it rules out animals taken from the wild and tamed (and hence not, technically speaking, 'domesticated'). Furthermore, while we can usefully apply the term to millions of animals in the West today, five hundred years ago it would only have pertained to, literally, a few thousand animals kept by aristocratic ladies.

We must rule out one misconception: 'pet' describes a role and not a type of animal. Nowadays many cats and dogs living in the West are companions pure and simple, and some of their previously useful attributes have become something of an embarrassment to their owners. The dog that barks at the gate and harasses those who try to enter the front garden performs a ritual once (in some locations, still) valuable to its owners. Sheepdogs may persist in trying to herd, switching their attention to children or cyclists unless trained not to. Cats' hunting instincts, so prized until quite recently, now inspire their owners with horror and many wildlife enthusiasts, keen to blame them (usually mistakenly) for declining numbers of songbirds and small mammals, with rampant disapproval.

For the purposes of this book, an operational definition of 'pet' may be the simplest. A pet is an animal that lives in the home (so not Aibo or other robotic 'pets'), is capable of moving around (not a house plant, however decorative), and is either constrained from leaving the premises (hamster, gerbil, snake) or voluntarily returns there (dog, cat). From an animal rights – and sometimes welfare – perspective, the latter may be an important distinction, but it is difficult to draw a hard-and-fast line, in terms of the emotional bond experienced by the owner, between a pet that lives in a cage and one that doesn't. The indoor-only cat – increasingly common in Western cities – highlights the arbitrary nature of such a distinction. Should we classify the indoor–outdoor cat with the dog, while putting the indoor-only cat into the same bag as the boa constrictor? Such distinctions seem misdirected. More important is the emotional complexity of the relationship, which is arguably richer with dogs and cats than with other species. Thus I will focus largely on these two species – or rather, their owners – for the remainder of this book.

DOGS AND CATS are so commonplace that it is easy to lose sight of the differences between them and also of the many changes that both have undergone over the past 10,000 years. Although dogs and cats share a common ancestor, they have always played distinct roles in human society and elicited different attitudes. Dogs, although often mistreated and occasionally eaten, have remained 'man's best friend' and are almost universally approved of; cats, on the other hand, we have long regarded as largely disposable. Attitudes towards cats have changed considerably from one era and one place to another. Religion and superstition intertwine with the domestic cat's history, and even today these animals can take on a symbolic importance. History records the ancient Egyptians as the first society to worship cats, a practice later taken up by pagan cults all around the Mediterranean. In the late Middle Ages, cats became associated with witchcraft, a tradition that persists in our celebration of Halloween. Even today, not all those who dislike cats hide their feelings. British controversialist Rod Liddle recently wrote in his *Sun* column, 'Are your kids sorting out anything special for Father's Day? Last year mine fixed for me to put down some cats at a local vet, an ambition realised after all these years.'[11]

At the other end of the spectrum are those whose enthusiasm for cats or dogs extends beyond their own pets: some devote themselves

The Egyptian goddess Bastet.

to caring for colonies of unowned animals (usually cats in the United States and United Kingdom but also dogs in other parts of the world). These rarely thrive on their own, and so not far from every colony there is usually a well-meaning person, stereotypically an elderly spinster but in reality more likely a middle-aged woman, who supplies food on a daily basis. Such caregivers must often ignore the objections of local residents, who complain about the mess the animals make and occasionally fear that they will transmit diseases. Thus many feeders operate off the radar. Those that work in the open speak of a powerful commitment to animal welfare and the deep sense of satisfaction they gain from taking care of 'their' animals.[12]

Thus the cat, perhaps more so than any other domestic pet, shows that the same animal can represent different things to different people. Even today, while many cats are simply companions, some people also cast them as healers, as useful predators, as hunters with cataclysmic prowess, as objects of extreme sympathy, or as production animals little distinct from pigs or sheep. Our attitudes towards animals are often complex and may also depend on the animal in question: for example, people who hate cats may be perfectly happy to own a pet dog (although Rod Liddle fantasizes about killing cats, he has a collie–Labrador cross called Jessie).

Even in the twenty-first century, when the divide between mere 'animal' and 'pet' has become sharper, the animals we loosely refer to as 'pets' may still be many things besides companions. They may

---

Attitudes Towards Animals[13]

The following lists different attitudes towards animals (I include only those relevant to species kept as pets):

Humanistic: interest in and strong affection for individual animals (i.e., pets)

Moralistic: concern over the treatment of animals; opposition to cruelty and the unnecessary exploitation of animals

Aesthetic: focus on the artistic qualities of animals

Symbolic: belief in animals' spiritual significance (e.g., as embodying human souls)

Utilitarian: use of animals for material commodities (e.g., meat, hide) or specific tasks (e.g., guarding, vermin control)

Dominionistic: appreciation of animals for the satisfaction derived from controlling them, often in sporting or competitive situations

Naturalistic: interest in and affection for wildlife and natural habitats

Negativistic: active avoidance of (certain) animals due to dislike or fear

These psychologically robust distinctions probably apply universally. Many people may apply several simultaneously to the same animal or situation, although a few seem incompatible, among American citizens at least. Not surprisingly, most people who express negative attitudes towards animals in general also disapprove of keeping pets, expressing incredulity that people could 'love' animals. Also, moralistic and utilitarian attitudes often clash over the methods used to exploit animals (e.g., hunting, 'factory' farming, laboratory experimentation).

---

be equally valued for their physical beauty (show dogs), for their usefulness (gun dogs), or for their ability in sports (agility champions). They may bring an element of wildness into the home (most outdoor-access cats). They may symbolize, or come to alter, their owners' stance on more general issues, such as the welfare of animals writ large. But all true pets have one thing in common: the affection of at least one person, who recognizes and values them for their qualities as individuals.

BEYOND THE EMOTIONAL dimensions of our relationship with animals, other attributes that contribute to pets' appeal at least deserve mention. We have and often still do value them for their beauty, which is both comforting and quintessentially other. French writer Colette, for instance, puts it beautifully with regard to *Felis catus*: 'The cat is the animal to whom the Creator gave the biggest eye, the softest fur, the most supremely delicate nostrils, a mobile ear, an unrivalled paw and a curved claw borrowed from the rose-tree.'[14] As visual creatures we humans pay a great deal of attention to what animals look like. Dog shows and cat shows are little more than beauty pageants, but the publicity they receive leaves many prospective owners believing they will only be happy with a dog or cat that looks a particular way. They seem blissfully unaware that the cuteness of many breeds comes at a genetic cost, not least because some features that appeal to us can entail a lifetime of discomfort for the animal.[15]

Pets can also confer status (or so their owners believe). Big cats and esoteric breeds of dog spring to mind, but large snakes probably epitomize this kind of pet. Snakes are generally so uninteractive that the relationship appears almost entirely one-sided, at least to the outside observer. However, they certainly mark their owner as unusual, even exotic. A man I recently met, who used to import snakes for the pets department at Harrods, claimed that girls used to respond with delight to invitations to visit his apartment for the purpose of meeting his enormous python. (After he married, his wife restricted him to dealing in spiders and scorpions.)

The prospect of an emotional bond, however, draws most people to the idea of owning a pet and sustains the relationship over time. Most owners are not only fond of their pets but believe their pets are fond of them. Whether this is actually the case depends in large part on the animal's species. People have long valued dogs especially for their apparently unconditional love for their owners: as Elizabeth Barrett Browning wrote of her cocker spaniel, Flush, 'His ears were often the first thing to catch my tears.'

Just because owners believe their pets love them, however, does not make it so. Modern biological thought – the 'selfish gene' theory – approaches such apparent altruism with scepticism, arguing that some advantage – contributing to the long-term reproductive success of either that animal or, failing that, its close relatives, who share many of its genes – must more than counterbalance genuine self-sacrifice by any animal.

Dogs are clearly not genetically related to their owners, so this 'unconditional love' requires some unpacking. *Canis familiaris* descended from a social ancestor, the wolf, which exhibits a degree of seeming self-sacrifice in the way the species organizes itself into packs. A pack starts with a pair-bond between a male and a female, whose offspring remain with their parents for one or two years to help raise the next litter or two of cubs – which are their full siblings and therefore share half of their genes. During this time they make no attempt to reproduce themselves, even though they are, biologically speaking, old enough to do so. Nevertheless, such altruistic behaviour is only temporary, since most eventually leave the pack to find their own mates. Moreover, while still living with their parents, they retain a measure of protection and also enhance their understanding of how best a wolf should inhabit its particular environment; juvenile wolves rarely have the cunning to thrive on their own. One theory of how wolves became dogs suggests that we have tapped into this system of family loyalty, such that during a puppy's second and third months of life, it comes to believe that humans are 'family', at least to the same extent, and probably more, than it accepts other dogs as such.[16]

While 'unconditional love' may be genuine (if genetically con-
trived) in dogs, this is less likely to be the case for other common
animal companions. Cats, descended from a solitary territorial ancestor, value place above all else. Hence moving home is a very different
experience, requiring quite different management, for dogs and cats.
Dogs are content to go where their owners go, but cats strenuously
attempt to return to the place where they used to live, seemingly
unable to deduce that since their owners are feeding them in the
new house, they are unlikely to perform the same service in the old
location. Nevertheless, it generally takes cats just a couple of weeks
to accept the new regime and abandon their quest to regain their old
stomping grounds. Woe betide cat owners who move only a short
distance and then allow their cats to escape: once cats find their way
back 'home', they can be very persistent in retracing their steps –
and may end up living with a former neighbour. Thus while cats
undoubtedly do love their owners, as shown by the rubbing, purring
and grooming rituals (the latter not always appreciated as such) that
they also use to cement friendly relationships with other cats, they
lay down quite strict conditions – food, shelter, protection – before
fixing their affections on people.

Other free-ranging pets, for example house rabbits and pot-bellied
pigs, may feel affection for their owners but possibly only as part of a
set of transactions based on food and shelter. Pets that we must confine to prevent them from wandering off, such as hamsters and mice,
self-evidently do not place a high priority on any relationship they
may have with the humans they find around them.

Reflecting the emphasis that most people place on the emotional
side of the owner–pet relationship, cats and dogs now have a much
higher status within Western families than ever before. Certain
unusual individuals throughout history have doted on very special
animals – eighteenth-century essayist Samuel Johnson fed his cats
Hodge and Lily oysters – but they have traditionally been regarded
as eccentrics. Nowadays, eight out of ten pet owners consider their
pets not just part of the family but as having equal status with the

human members. Over three-quarters of US pets enjoy equivalent status to children. As our lifespans continue to increase and the size of the average household shrinks, more people find themselves living alone or as childless couples: pets seem to be filling the void, serving as beings to care for. The term 'pet owner', which to some implies a politically incorrect degree of dominion, is giving way in certain quarters to 'pet parent' or 'pet guardian'.[17]

The emotional side of the relationship often runs to more than one-way affection. For some, pets can seem to provide much more, a substitute for the ties with other people that are increasingly difficult to sustain. I suspect that Caitlin Moran, award-winning columnist for *The Times*, would support this notion. She admits to 'hating' her cats when her husband and children are at home, but within twenty minutes of their leaving, she writes, '[I] will hunt through the house to find the cats, and drape them over my shoulders. I dote on them.' So, for her at least, cats become a fitting substitute for people when there are no suitable humans around. Nor have Ms Moran's cats inhibited her from having children. Some have accused 'Generation Y' of doting on their pets so much that they never get around to producing offspring of their own.[18]

RECENTLY, MANY PEOPLE have put forward claims that the significance of pets runs deeper than simple emotional gratification. Those wishing to promote pet keeping seem to crave a greater justification than the simple happiness that (most) owners derive from their pets. Many point to research indicating that animal companions are a panacea for the stresses of modern life. They suggest that the company of an animal might do more than just make us feel good; pets' supposed effects on cardiovascular health can potentially extend our lifespans. The idea of companion animals (mostly dogs) as therapists, once dismissed as the raving of a lunatic fringe, is now so widely accepted as to be scarcely newsworthy, even for editors who know that articles about pets sell papers. In fact, the original concept of therapy dogs now extends to other species, such as cats – even pigs.[19]

This new conception of pets as universal life enhancers has spread remarkably quickly, despite decidedly patchy scientific evidence. Ironically, although the idea of pets as therapists began in a psychotherapist's office and the suggestion that pets might enhance cardiovascular health began with a research study by a professor of health sciences, science has struggled to keep pace with the enormous swing in public opinion regarding pets. Led by an enthusiastic press, most people in the contemporary West seem only too happy to accept the idea that pet keeping is much more than a simple pastime. Popular culture now promotes pets widely as calming (though try telling that to the owner of a disobedient dog in the park), life-extending, and capable of bringing even the most disabled or disturbed children out of their isolation.

We may seek some parallels in the growing popularity of alternative and complementary medicines during roughly the same period. For example, homeopathy has gained wide acceptance in Europe. The homeopathy industry is currently worth well over €1 billion annually, despite the scientific viewpoint that the majority of homeopathic remedies contain no active ingredients whatsoever and work by a combination of suggestion and placebo effect.[20] Unlike with homeopathy, at least some science supports the healing powers of pets, but this is unlikely to account for the idea's rapid and widespread acceptance. After all, those who take homeopathic medicines seem happy to ignore the scientific evidence, so why should those who believe that pets can improve their health take much notice of science, whether or not it supports their case?

Perhaps a more interesting question is, why, in an age of supposed reason, have a few kinds of domestic animals become imbued with almost mystical powers? With the benefit of hindsight, we can understand why our less well-informed ancestors might have believed that some animals had magical powers – for example, the 'spirit friends' that appeared to Native Americans when they entered trancelike states – or, conversely, held them responsible for calamities of various kinds; consider the dogs and cats exterminated in London in 1666 based on the (mistaken) belief that they carried bubonic plague. And yet, it seems we are no different, even though the idea that some animals have evil intent while others actively intervene on our behalf has given way to the notion that all are generally benign, if unwittingly so. What in human nature wants so desperately to believe in the healing powers of pets?

NOT EVERYONE IS EQUALLY susceptible to the charms of pets. Many people are indifferent to them, and others find them repellent. Some portray pet keeping as harmful to the environment. Because dogs and especially cats need to eat meat, and thus their maintenance contributes to unsustainable agricultural practices, their environmental impact has been likened to that of an SUV. Cats 'murder' songbirds and cute small mammals. Horses take up pasture that could grow crops for human consumption. Saltwater-fish keepers stand accused of raiding fish populations on endangered coral reefs. Perhaps to

About one person in twenty is cat phobic.

counteract the positive propaganda that pets have received, such objections seem to be intensifying.[21]

Some critics of pet keeping ask whether such a degree of emotional investment can be healthy, either for the individual concerned
or for society as a whole. Just as they assert that the money spent on
pets would better serve humans, so they question whether emotion
squandered on mere animals (who in their view do not appreciate
such attention) diminishes pet owners' consideration for other human beings. They charge, essentially, that there is a finite amount of
compassion in the world, and animals are consuming more than their
fair share. From this perspective, rather than a boon to society, pets
are more like a cancer, a growth that sucks resources from the needy
and provides only imaginary gratification in return.[22]

Given the chance, pet enthusiasts fight back. When in February
2016 social commentator Laura Marcus stated in an opinion piece on
the Guardian's website, 'Your pet doesn't love you – it's just trapped
by you: I find it easier to explain not having children than my dislike
of keeping pets', readers bombarded her with vitriol. The moderator
approved comments like 'Blasphemy!' 'If you think of *any* animal
as an "it" there's something wrong with you.' 'What a sad commen-
tary . . . Choosing not to have relationships is to choose emotional
poverty.' The passion with which many people defend their choice
to keep pets shows just how deeply embedded these companions
have become in our psyches.

Marcus, however, was simply voicing a personal perspective.
She was not exhorting others to abandon their cherished pets. Nor
was she expressing a general dislike of animals: on the contrary, she
started her piece with the words 'I love animals, I just don't feel the
need to own them.'

Why do arguments over whether it's OK to love animals get so
heated? It's because, for many of us, our animals have become an extension of ourselves, an emotionally based form of self-enhancement.
Owners regard their pets as reflections of their personalities, and
canny politicians are fully aware of this. In 2016 Marine Le Pen,

leader of France's 'historically brutal' Far Right National Front party, posted pictures of her pet cats and kittens on her blog in an apparent attempt to portray herself as 'a soft, sensitive soul'.[23] It has to be done right: Mitt Romney's chances as a candidate for the US presidency disappeared with the revelation that he'd once taken his dog on holiday with him in a crate strapped to the roof of his car.

Surely affection prompts owners to leap to their pets' defence, and also primes them to believe that keeping an animal in the home is anything but frivolous. Pets are part of the family, and an attack on one can feel like criticism of a close relative. Contemporary trends cannot possibly account for such deep-seated feelings.

PET KEEPING IS by no means a modern affectation. A considerable body of evidence indicates that affection for animals is an intrinsic part of human nature that has shaped both who we are today and how we got here.

Our interactions with animals have shaped our mental as well as our physical evolution. Our brains differ from those of other mammals in our ambition to form some idea of what other animals might be thinking. This ability must have originally evolved because it enhanced our effectiveness as hunters. However, it evidently came with an unexpected twist: the respect that today's remnant hunter-gatherers express for their quarry suggests that such comprehension automatically carries with it a sense not of contemptuous superiority but of affection, which elevated our predator–prey relationships above unalloyed exploitation.[24] Not long after they developed that sense of camaraderie with wild beasts, our distant ancestors started bringing individual animals home and taking care of them. Thus began the practice that eventually evolved into modern pet keeping.

Over the thousands of years that followed, humans continued to keep animals as companions, the focus of their affections narrowing as a small selection of species became domesticated and thus more adaptable. Fashion and economics also played a role, restricting out-and-out pet keeping to those few individuals with sufficient leisure and the resources to indulge it – although affection for working animals, especially dogs and horses, seems to have been widespread. The hankering to build close relationships with animals evidently runs throughout human history, albeit in many different forms. Surely it is not too fanciful to regard this as a universal trait of mankind.

WE CRAVE A CONNECTION not just to animals but to nature in general. Urbanization has taken most of us out of the wild but evidently has not eradicated our yearning for it. The animals (and plants) we keep in our homes, the pleasure that many of us obtain from tending gardens and visiting wilderness locations, all speak to an urge to engage with the natural world.

Once the human species' very survival depended on an intimate knowledge of the plants and animals that surrounded us: nowadays, like so much else, we delegate that responsibility to a few specialists. We in the West take for granted that the food we eat will be wholesome, safe and readily available. Horror strikes when so much as a mouse strays into our homes. Our lives are becoming so sanitized that our health, rather than improving as we isolate ourselves from germs, is actually suffering from a lack of interaction with other species – in this case, with microbes.

Yet, even as our lifestyles increasingly distance us from nature, we find ourselves spending a great deal of time and money trying to reconnect with it. The cut-flower industry generates about $7 billion each year in the United States; in the United Kingdom, the combined sales of flowers and indoor plants reach an estimated £2.2 billion annually (bringing in slightly more than the United Kingdom's much-lauded music industry). In continental Europe, per capita spending on flowers and plants for the house is at least double that in the United Kingdom. This despite the ever-improving quality of artificial alternatives, which we somehow still regard as inferior to their natural counterparts.

Many of us also evidently feel the urge to reconnect with animals. Our spending habits betray how important this can be. In the

2011–12 winter the British Trust for Ornithology reported that, despite pressures on household budgets, spending on food for garden birds remained high, especially among people over sixty-five. More than 40 per cent of US households practise garden bird-feeding, an activity even more popular in the United Kingdom. Many people evidently choose not to live without the enjoyment they get from wild birds.[25]

However disconnected from the natural world they may seem, our pets are ultimately animals. Sentimentality aside, the ways we think about and relate to them arise out of our evolutionary journey from hominid to *Homo sapiens*. Fundamentally, we are highly evolved primates, carrying an evolutionary legacy that affects everything we think, feel and do. Were our behaviour infinitely flexible, we should be willing to use our technologies to finalize our ongoing divorce from nature. And yet we seem curiously reluctant to do so, implying that interaction with plants and animals satisfies psychological needs that defy pure logic. Thus we can view pet keeping in the wider context of our perhaps irrational but nonetheless powerful urge to connect with the natural world.

Pet keeping is an ancient tradition that has evolved into a modern custom. It has involved many of the same types of animals – dogs, cats, birds, even pigs – throughout its history. I believe that our desire to fill our houses with animals, when we have little practical need to do so, is one way of expressing what it means to be human. Animals have played a major role in our evolution, first, in the hunter-gatherer era, as important sources of protein, and then, after domestication, as both food and currency. Culturally, animals have always been important to us: the earliest cave paintings feature far more beasts than humans, and anthropomorphized animals dominate today's children's stories. All this suggests that our interactions with animals may have shaped the structure of our brain – the organ that distinguishes us most from all other mammals. If this is true, pet keeping may be a habit that our basic humanity will not allow us to give up.

Cats and Warfare

Historically, the cat's reputation as an associate of Satan must have justified all manner of callous treatment, much of it unrecorded. A manuscript recently unearthed at Heidelberg University describes a bizarre proposal for the use of cats during the siege of a fortified town: Franz Helm, a sixteenth-century artillery specialist, advised the attackers to 'obtain a cat from that place' and bind to its back 'a small sack like a fire arrow', then 'let the cat go, so it runs to the castle or town, and out of fear thinks to hide itself. Where it ends up in barn hay or straw it will be ignited.' Helm had evidently acquired an understanding of cats' behaviour while remaining entirely indifferent to their welfare.[26]

CHAPTER 1

Pets, Ancient . . .

W E MAY NEVER KNOW when our ancestors became more than just hunters and began to form lasting, personal relationships with individual animals. The archaeological record can take us back only 12,000 years, to the first known burial of a dog with its master, in what is now northern Israel. But this long-deceased canine was surely not the first pet, even though dogs were the only domesticated animals at that time. Much earlier, if the habits of twentieth-century hunter-gatherer societies are anything to go by, our distant ancestors took animals of many kinds from the wild and raised them to live with people. And so our ability to understand and feel affection for animals must be an ancient trait that most likely emerged as our brains evolved into their current form, about 50,000 to 30,000 years ago. If this is correct, pet keeping makes up an intrinsic part of what it is to be human, and so we should be able to trace some common thread in the practice that transcends both time and culture. The recorded history of pet keeping, however, can only tell us half the story. As the mists of time obscure its origins, in seeking its essence I will initially concentrate on the many similarities between relationships with individual animals in otherwise very different contemporary cultures, ranging from the relict hunter-gatherers of Amazonia to the socialites of Manhattan.[1]

We have no direct evidence proving that people living prior to 10,000 BCE had pets. Any kept by hunter-gatherers must have included species tamed from the wild, which would leave little

23

archaeological evidence: their remains would be impossible to distinguish from those of animals killed for food or kept for other – perhaps
ritualistic – purposes.

Since we don't have evidence from the prehistoric past, we must
look to that gleaned from the past century. A remarkable number of
hunter-gatherer and small-scale horticultural societies that persisted
into the nineteenth and twentieth centuries in remote parts of the
world – Amazonia, New Guinea, the Arctic and elsewhere – give
us insight into the behaviours of earlier Stone Age societies. We can
start by asking whether hunter-gatherers already kept pets when they
were first documented, before they had time to acquire the habit from
the West. If they did, then we will at least know that pet keeping is
compatible with such a lifestyle and so may assume with reasonable
confidence that our pre-agricultural ancestors kept them too.

It turns out that many small-scale 'Paleolithic' societies kept pets
of some kind: sometimes dogs, but mostly tamed wild animals, captured when young and then brought up as part of the human family.
Native Americans and the Ainu of northern Japan kept bear cubs;
the Inuit, wolf cubs; the Cochimi from Baja California, racoons; indigenous Amazonian societies, tapir, agouti, coati, and many types
of monkeys; the Muisca of Colombia, ocelots and margays (two local species of wild cat); the Yagua of Peru, sloths; the Dinka of the
Sudan, hyenas and monkeys; native Fijians, flying foxes and lizards;
the Penan of Borneo, sun bears and gibbons.

To this list of pets we can add a host of bird species, valued as pets
from Brazil to Mali to China. Many have particularly bright plum-
age, such as parrots, parakeets and hornbills; others, such as the bul-
bul, sing. Selection of some – such as the cassowaries, large flightless
birds, cherished by the original inhabitants of New Britain (part of
New Guinea), and the pigeons kept as pets in Samoa – seems to have
been more arbitrary. Nowadays, the availability of Western domesticated animals has reduced some of this diversity, but traditional
societies, from the Toraja in the mountains of Indonesia to the Tiv of
West Africa, still widely treasure animal companions.[2]

While traditional cultures do keep an extraordinarily wide variety of animals, a recent survey of sixty such societies finds that dogs and cats are nonetheless the most ubiquitous. This preference is clearly not traditional in most cases, since dogs and cats arrived in most parts of the world very late. Dogs were almost certainly domesticated (from the Eurasian wolf) by one or possibly several hunter-gatherer societies several thousand years before the dawn of agriculture and then gradually spread throughout much of the globe. However, dogs were unknown in some parts of South America (for example) until their arrival with European explorers, at which point they quickly displaced the local 'Aguara dog', a much less trainable domesticated fox. Dog keeping may thus reach back more than ten millennia in some societies, such as among the Saami of Finland, but only a couple of hundred years in others. Cats, domesticated less than 10,000 years ago in the Middle East, are less versatile and would have made their appearance more recently than dogs in most parts of the world. The greater prevalence of dogs in the traditional societies surveyed reflects this chronology. Dogs appeared in fifty-three of the sixty societies surveyed, while cats appear in only about half.

Because both dogs and cats have practical uses besides companionship, their status is not always easy to determine, given the cultural and linguistic barriers that often exist between Western researchers and traditional peoples. The survey found that about one-third of the groups in which dogs occurred treated them as pets; another third did not regard them with affection but simply used them as guards or for some kind of work. As expected, those groups that had cats regarded them as useful for the control of vermin, and two out of three such societies thus expected them to find their own food. In the others, however, certain individuals ('owners') deliberately fed at least some of the cats and treated them as pets. The same pattern emerges with other domestic species. About one culture in ten kept either pigs or horses or another type of hoofed mammal (cattle, camels, water buffalo, sheep) as pets; of course, many more kept these species purely as livestock.

While widespread in these traditional societies, cats, dogs and other familiar domestic animals represent only a minority of the vast range of species kept as pets. The survey recorded many kinds of tamed mammal, including primates of various kinds, foxes, bears, prairie dogs and ground squirrels. Over a quarter of the societies also kept birds, which were even more varied than the mammals, including eagles, ravens, parrots, macaws, hawks and pigeons. Although evidently valued for their appearance, most of the bird species kept had higher than average intelligence (for birds). Many clearly formed lasting relationships with humans: for example, the Yanomamo of South America taught their parrots to talk. Overall, the birds more obviously served purely as pets than most of the mammals: almost all received most of the food they needed, and many functioned as playthings for children.

Fish are the only class of pets almost entirely missing from traditional societies, presumably because appreciating them requires glass for aquaria. An exception: the Polynesians of Samoa captured and then tamed eels, keeping them in holes in the ground and whistling to call them to the surface (given the current popularity of garden ponds, a revival of this practice might be amusing).

Overall it seems that most of these societies are (or, until very recently, were) much more physically integrated with animals than we are today. This is hardly surprising, given that many still depend for their livelihoods on a deep understanding of the behaviour of the animals around them. But remarkably, the relationships between humans and animals in these societies very often extend far beyond the purely utilitarian. They treat most of the multitude of animals kept in the home with great consideration and kindness.

It is, of course, tempting for us to view these human–animal relationships through the lens of our own 21st-century pet culture and to presume a continuity between pet keeping 'then' (in modern 'Paleolithic' societies) and now (in the industrialized West). But while the human–animal relationships within such traditions may look strikingly familiar to us, they also exhibit some deep dissimilarities.

To EXAMINE THE DIFFERENCES between modern pets and ancient animal companions, we need to turn away from generalities and examine individual instances. Among the best studied are the Guajá, a hunter-gatherer society in the northeastern Brazilian state of Maranhão notorious for keeping large numbers of pet monkeys – a tradition its members have been reluctant to abandon, even though they now know that the monkeys act as reservoirs for tuberculosis and therefore pose a real threat to human health.

The origins of this tradition are not easy to determine, given the convoluted recent history of the Guajá. Their current lifestyle is probably the consequence, direct or indirect, of European colonization. (In this they are not unique: many of the so-called primitive societies extant today survived by migrating away from the colonizers, thereby avoiding enslavement or decimation by Western diseases, such as cholera and smallpox, to which they had no natural resistance.) In the eighteenth century the Guajá were probably horticulturalists, living in areas now occupied by other tribes and people of European descent. They may even have been enslaved by the Portuguese for a time, before migrating eastward and of necessity reverting to a nomadic lifestyle. They obtain much of their protein by hunting, living off wild peccaries (the New World equivalent of the Eurasian wild boar) year-round, fish in the dry season, and monkeys, chiefly the howler monkey, in the wet season.

The men collect their monkey pets during the hunt. When they kill an adult female – a howler or, less commonly, a black-bearded saki, an owl monkey or a tamarin – they try to capture any dependent infant and bring it back to the village. There, they hand the monkey over to one of the women, who decides whether to kill and eat it or to keep it as a pet. If she chooses the latter, the little monkey receives diligent care, after which it becomes taboo to harm or kill and eat it, even if, as often happens, it injures people once it has grown to adulthood. The Guajá word for 'pet monkey' resembles that for 'child' and is quite distinct from that used to describe other pets, which include coatis, agoutis, peccaries and tortoises (they also keep domestic dogs, but although valued as guards, these supposedly 'lack souls' and often endure cruel treatment).

Most remarkably, the younger women in the tribe breast-feed the infant monkeys, often alongside their own babies. As the young monkeys mature, the women premasticate their food, allowing them to grab it from their mouths. Guajá women carry the monkeys everywhere, and a given family's most senior female member may walk around with three or four young monkeys entwined around her shoulders and head.

The Guajás' devotion to monkeys comes at a significant cost. Typically, the number of monkeys in a given village will roughly equal the number of humans, so they consume a significant amount of food. Moreover, as they get older, most become heavy and awkward to carry – especially the howlers, which can weigh over ten kilos. They can become aggressive, possibly a consequence of poor socialization due to their orphaning. Thus Guajá women rarely carry adult monkeys, which they keep tied up. Sooner or later, most escape and return to the wild.[3]

How to characterize the relationship between a Guajá woman and her monkeys? While intense, to an extent unsurpassed in the West, the bond is less powerful in other ways: notably, it does not last for the animal's whole life. Most modern dog owners, for instance, wouldn't dream of abandoning Fido as soon as he reached adulthood. But Guajá women banish the monkeys when their cute, portable, affectionate pets grow into heavy, grouchy adults. The relationship differs in other ways too. It is explicitly gendered: while men may keep other pets, they rarely adopt monkeys. The monkeys also seem to serve a kind of decorative function for the women, who 'wear' them, almost like a form of body art. And it seems to have a competitive component, since the most senior women – the Guajá are a matriarchal society – tend to carry the most monkeys.

So what function do the monkeys fulfil? In one sense, the relationship is not utilitarian: wild monkeys are eaten regularly, but it is taboo to eat a pet monkey. Yet in another, it may be: the young monkeys may play an educational role. Growing up with monkeys of several species, boys learn about the habits and behaviours of an important prey item. Boys receive small bows and arrows while still very young and, despite the proscription against killing them, use pet monkeys for target practice (child rearing among the Guajá is, by our standards, remarkably laissez-faire). Post-pubescent girls receive monkeys to care for before having children of their own, perhaps to inculcate valuable child-rearing skills.

The Guajá seem to place baby monkeys in the same category as their own children, anthropomorphizing them to an even greater extent than we do our 'handbag' dogs. For example, they feed their pet monkeys on the same diet that they eat, including monkey meat, even though the Guajá know perfectly well that this is not the monkey's natural diet. Their dogs, by contrast, fed whatever is to hand, must constantly scrounge for food, including scraps dropped by the monkeys. In this, the Guajá are a lot like us in prioritizing one species over most others. We'll feed pigeons and squirrels breadcrumbs now and then, but we save the filet mignon for our cocker spaniels.

To OUR EYES, the most unexpected, even unsavory, aspect of the Guajá women's relationship with their monkeys is the breast-feeding. Yet far from an aberration confined to one tribe or just to monkeys (as the most humanlike of animals), breast-feeding of pets used to occur in small-scale societies all over the world and still does in a few, as evidenced by many twentieth-century records from isolated societies in and around Oceania, especially the Malay Peninsula, New Guinea, Melanesia, Micronesia and Polynesia. Breast-feeding animals, including monkeys and peccaries, was also widespread in South America within the past century, while the Maya suckled deer fawns.[4]

We can never be entirely sure how ancient the practice of breast-feeding animals is, but there is no reason to suppose that it was not widespread 30,000 years ago. In some instances it appears to stem from a long tradition and not to involve any one animal type. For example, Maori women, already breast-feeding puppies, took to suckling piglets within fifty years of pigs' introduction to New Zealand by European settlers in the mid-nineteenth century.

In most cultures where humans breast-fed animals, they had usually obtained orphaned young animals during hunting expeditions. In a few cases, they bred the animals specially. In New Guinea, the nursing of piglets used to be almost ubiquitous, but different tribes obtained their piglets in different ways. In the lowlands, villages did not breed pigs deliberately but castrated all tame boars and obtained almost all piglets from the bush. The absence of lactating sows required that piglets be breast-fed. In the highlands, villages generally kept one or more stud boars and bred the piglets, which women suckled only if the mother pig lacked milk or died.

Pigs are by no means the only species breast-fed then eaten by humans. In northern Japan, the Ainu, originally a hunter-gatherer society, have a special relationship with the local Ussuri brown bears, likely ancestors of the American grizzly. In their religion, Kim-un Kamuy, the god of bears, is one of three important deities. Until the mid-twentieth century bear hunting was a socially significant activity, and bear meat is still regarded as a delicacy. The Ainu used to hunt bears as they came out of hibernation, when they had not yet fed and regained their strength and were easier to catch. The tribesmen killed the adult bears immediately and feasted on their meat but took any captured cubs back to the village, where the women nursed them and fed them premasticated food (see nearby illustration). As they grew, the young bears became difficult to control and were therefore kept in cages. This may seem cruel to our modern sensibilities, but they were well fed, consuming as much food as all the human members of the family put together. Then, when the bear was older – between one and three – the Ainu ritually sacrificed it and displayed its skull atop a wooden post in the ground. At these ceremonies, the woman who had raised the bear from a cub displayed considerable grief, even attacking the bear's killers with a tree branch. It is unclear how much of her emotion was genuine and how much was ritual, since later in the event she might join in the dances that followed the killing.

While the Guajá prohibit the killing of their breast-fed monkeys, other cultures do not follow the same proscription. The native

Ainu woman suckling a bear cub.

Polynesians of Hawaii, who breast-fed dogs until at least a hundred years ago, sometimes used the animals as ritual scapegoats: the mother of a child who had died would kill a dog she had suckled as a puppy, and the two would share a grave. Elsewhere in Polynesia, people routinely ate breast-fed puppies, reputed to produce the tastiest meat.

Small-scale societies most commonly breast-fed dogs, likely because they were the animal companions most widespread among them, but the cultural context and the eventual fate of the animal varied from one society to another. The Seri of northwestern Mexico and the Arapaho of Wyoming only allowed puppies to suckle from women producing too much milk, apparently believing that this improved the quality of the milk taken by the woman's baby. The Onge of the Andaman Islands suckled dogs routinely but exclusively (never nursing the young of wild animals they collected) and showed great affection for these hand-raised pups, which they valued both as companions and as hunting aides. Tradition among both the Tupinamba of Brazil and the Arawak of Suriname holds that breast-fed dogs grow into superior hunting companions to those nursed by their mothers. In Australia, Aboriginal hunters used to collect dingo puppies and suckle them (the dingo is a feral dog, descended from domestic dogs that went wild some 3,000 years ago). They rarely if ever ate these puppies, though they might other breast-fed pets. The dingo pups served as pets for the children and as living bed warmers for the adults until they became a nuisance, when they were encouraged to go back to the wild. Thus the specific culture determined the significance of a suckled puppy: some ate breast-fed dogs, some sacrificed them, some trained them as useful working dogs, and others adored them as pets for the whole of their lives.

All this demonstrates that while breast-feeding animals seems astonishingly intimate to us, it doesn't always generate deep emotional attachment. An underlying nutritional factor – the availability of alternative foods suitable for unweaned mammals – likely came into play. We have little historical evidence of women breast-feeding animals in places with wide access to milk from domesticated animals – Europe and Africa, for instance. There, orphaned mammals, whether raised for food, ritual purposes, or as pets, could drink milk obtained from farmed animals. In places without wide access to animal milk until comparatively recently – for instance, the Americas and the islands of Southeast Asia – breast-feeding provided nutrition for animals that would otherwise have perished.

In most cases, a woman nursing a young animal pays a nutritional cost. What might the subsequent pay-off be? Once trained, a well-socialized hunting dog, breast-fed as a puppy, might help a woman's husband acquire a great deal of food. But how do we explain customs like those of the Guajá, who don't eat the monkeys they collect as babies and only seem to bond with them during the early years of the animals' lives? In such cases, perhaps breast-feeding animals enhances a woman's status – and thereby her and her family's chances of reproductive success. The Ainu family providing the bear at each ritual feast rises in status for the whole subsequent year. Among the Guajá, power resides with the grandmother who carries the most monkeys.

Because Western societies encourage women to breast-feed their own infants only in private, human suckling of animals seems so extreme that we risk assigning it too much importance. We must guard against projecting modern significations onto ancient practices. Just because women in other cultures interacted with animals in ways that seem unfathomably intense to us does not mean that they automatically considered them 'pets' in the sense that we use the word. Very few of us can imagine breast-feeding Spot the dog, and even fewer could countenance boiling him in a pot and serving him for dinner. But both stances reflect cultural norms: while breast-feeding of puppies is nowadays almost unheard of, in the Far East dogs are still bred specifically for meat.

Even when other societies' behaviours resemble our own, we cannot assume equivalence. People in very different cultures who treat animals with considerable affection are not necessarily 'just like us'. Their motivations may differ vastly from our own. For example, while some native Amazonians abhor the very thought of eating an animal that has become a 'pet', the ethical background to their attitude may vary quite a bit from that underlying our own taboos on killing and eating pets. Although pet keeping in small-scale societies may include practices that conflict with our modern sensibilities, its relevance to modern pet keeping lies in the nature of the emotional bond between animal and owner.

HUNTER-GATHERERS HAVE a much more complex relationship with animals than we in the West do today. In their societies, animals serve both an essential function (as food) and a symbolic one. For example, caring for an orphaned baby animal may represent atonement for the harm done to its kin through the hunt. The Huaorani, an Amazonian people living in Ecuador, adopt baby monkeys and other jungle animals, just like the Guajá some 3,000 miles distant. When hunting adult monkeys, they use blowguns to shoot poison-tipped darts, then attribute the animal's death not to the dart but to the plant from which they extracted the poison, as if to distance themselves from the deed. After killing a female monkey, they attempt to capture any young still dependent upon her. Orphaned baby monkeys and other young mammals and birds brought home from hunting expeditions are not merely tamed; they are adopted as mem- bers of an extended family with both human and animal members and fed choice fruits. Tame harpy eagles partake of the meat from adult monkeys killed in hunting expeditions. On their deaths, these animals receive ceremonial burials.[5]

Such rituals occurred in the Paleolithic era. In one particularly strik- ing example from 16,000 years ago in Jordan, archaeologists discovered the skull of a red fox buried next to the remains of a woman on top of a layer of red ochre, a pigment of special value and ritual significance. Even more remarkably, this was most likely to have been reburial of both human and fox, since some parts of their skeletons remain in an- other grave close by. Whoever moved the body of the woman appar- ently knew of her relationship with the fox and moved the animal's most obvious remains – notably the skull – with her. It seems implau- sible that the fox died coincidentally at the same time as the woman; rather, it was almost certainly killed when she died, presumably so it could accompany her to the afterlife. Though we may deem this un- necessarily cruel, it does indicate a very special relationship between the two. We will never know precisely what that relationship was, but as no evidence exists of domestication of foxes there or anywhere else for another 12,000 years, we can reasonably assume that this fox had been obtained from the wild as an unweaned cub, in the style of other Paleolithic pets. Red foxes, tamed or otherwise, have no known use in hunter-gatherer societies except as sources of fur, so this individual was likely no more and no less than a much-loved companion.[6]

Beliefs about animals form an essential part of the spiritual life of such small-scale societies. The role of the animal, and therefore its treatment, can vary widely, depending on a given society's tra- ditions. For example, in Amazonia, the Aché make pets of coatis (group-living racoon-like carnivores), believing that their wild rela- tives transport human souls to the land of the dead. By contrast, the neighbouring Arawete believe that coatis feed on human corpses. Not only do they not keep them as pets, they set fires around newly dug graves to drive any nearby coatis away.[7]

Thus while all close personal relationships with animals must entail a degree of affection, other factors drive pet keeping in extant 'Paleolithic' societies. Not only are animals spiritually important, but their possession may convey more tangible benefits in the form of status and prestige. Indeed, in the absence of taboos, animals treated like pets when food is plentiful may become a convenient 'walking larder' of protein when hunting fails: the Campa in Peru and the Yupa in Colombia both consume former pets in non-ceremonial contexts.[8]

Extant hunter-gatherer societies' diverse relationships with animals point to an equally complex picture 30,000 years ago in the late Paleolithic, before the actual domestication of any animals. By that time the relationship between humans and the animals around them had very likely already advanced far beyond that of hunter and prey. Most such societies would have been multi-species groups, comprising human families, each with a selection of young animals obtained from the wild and raised alongside the children. Because these societies had no domesticated milk-producing animals, breast-feeding of orphaned mammals was probably more or less ubiquitous. Although we can never be certain, it is difficult to imagine how these relationships could have been sustained without feelings of deep affection for the individual animal concerned, despite the likelihood that it would end up on the dinner table or be driven back into the wild once it had become more nuisance than companion.

Thus our species' capacity to form an affectionate relationship with an individual animal quite probably stretches back to the beginning of the Upper Paleolithic, some 50,000 years ago. That era pre-dates the domestication of any animals, so pet keepers must have obtained all their animals in the wild, unlike their modern equivalents in Amazonia and New Guinea, who have found imported domestic dogs, cats and pigs easier to keep than wild-caught pets. Nevertheless, as recently as a century ago, alongside their puppies, kittens and piglets, almost all still kept as pets animals traditionally adopted from the wild – evidence that while pet choices increasingly reflected options recently imported from elsewhere, pet keeping itself is quintessentially ancient.

PET KEEPING AMONG modern hunter-gatherers also demonstrates the impact of domestication. While the practice began with wild animals, it evolved in such a way that those animals changed their natures to become more compatible with their owners' practical and emotional needs. The key transition would have come when some Paleolithic pets began to breed in captivity. Previously, there would have been no mechanism whereby captivity could alter the animals' genetics. Undoubtedly they would be much easier to tame and socialize (generally speaking, the younger the wild animal when taken from its mother, the stronger the bond it develops with its captor, so breast-fed animals would often have been more tractable than those caught soon after weaning). The mere experience of living with people would not have affected these early pets' genetics – which instead promoted the ability to survive and reproduce in the wild – and so would have left no impression on the species in general. Breeding within the human–animal 'family', on the other hand, would have provided the potential for genetic change. We need not invoke deliberate attempts to produce a better pet on the part of the humans concerned. Progress towards domestication would simply have entailed a tendency among those individual animals more suited to living with people to stay and breed, while the less tractable returned to the wild, either of their own volition or with the encouragement of their now less-than-enthusiastic owners. Animals born in proximity to people can adapt to life in a mixed-species society from day one; every individual need not be tamed from scratch. In this context, natural selection favours those animals best suited to breeding alongside humans, gradually changing their temperament to fit better with human requirements.

Accumulated over thousands of generations, such changes can be extremely powerful. Domestic dogs' brains contain irreversible differences from those of their ancestors, various Eurasian wolves, so that they can almost effortlessly learn how to interpret and react to human behaviour. Domestic cats, while retaining an air of inscruta-bility, are nevertheless a long way removed from their wildcat ances-tors, which are not only almost impossible to tame but which are also entirely unable to coexist alongside other members of their own kind, much less other species.

Although domestication of animals (and plants) is a defining characteristic of the transition from the Paleolithic to the Neolithic some 10,000 years ago, the domestication of the dog from the wolf occurred much earlier, when all humans lived as hunter-gatherers of one kind or another. We kept dogs before we invented writing, before we had permanent homes, before we grew crops. The precise details of when and where such domestication first occurred are still a matter of debate. Archaeologists have found remains of doglike animals dating from around 30,000 years ago in both Belgium and the Czech Republic, though we can't at present tell whether these were the direct ancestors of today's pooches, a semi-domesticated species that died out, or an unusual kind of wolf that no longer exists. Some scientists claim that the domestic dog originated not in Europe but in Siberia or southern China. Maybe wolves became domesti-cated in several locations, later interbreeding with gusto as soon as their human masters had established international trade routes many thousands of years later, and thereby confusing the evidence.[9]

The path from wild wolf to domestic dog, the first domestic species, cannot have been straightforward and was probably not deliberate: no precedent would have existed for the idea that a wild animal could reproduce in captivity. The current prevailing theory holds that dogs domesticated themselves, descending from an unusual type of wolf that no longer exists in the wild. These wolves would have differed from their wary modern-day counterparts in being sufficiently toler-ant of humans to spend much of their time scavenging around their camps. Adopting their cubs as pets would then have been easy, begin-ning the process of genetic selection towards tameness and, eventu-ally, trainability. Initially these early 'proto-dogs' might have served as early-warning devices (accounting for why dogs are much more prone than wolves to bark) and possibly as waste disposals. Not until they had become capable of forming social bonds with humans would they have been sufficiently controllable for useful service on hunting expeditions. Thus, although utility would have provided the usual motivation for keeping a dog in the Paleolithic, that usefulness would have stemmed from the ties of affection. The slavering, perpetually chained guard dog aside, dogs' effectiveness stems from their attentiveness to and desire to please people. The loved puppy that has become well adjusted to the humans with whom it will spend the rest of its life will be the most attentive and easiest to train. Unlike with most of the other domestic animals that came along later, the relationship between dog and master is fundamentally an emotional one.

The enormous number of dog burials unearthed from the period between 14,000 and 4,000 years ago indicates the esteem pre-literate societies had for dogs. One of the earliest and perhaps best known comes from the upper Jordan Valley, where archaeologists discovered the skeleton of an elderly human (although both skeletons were mostly well preserved, the pelvis of the human was too badly damaged to determine its sex), with its hand resting on the chest of a puppy, which may have been killed for burial. Dating to some 12,000 years ago, the culture that buried this pair, the Natufians, were on the cusp of the transition from hunter-gatherers to settled agriculturalists. The positioning of the two skeletons strongly suggests a very close and affectionate relationship between the human and the animal – as if the puppy was intended to accompany its owner into the afterlife.

Because dogs can be servants as well as companions, we cannot automatically assume that a dog found buried intact was a pet. Despite documentation of ancient dog burials all over the world, the motivations behind them seem to vary from one place and time to another. In many graves the interred dog has been carefully positioned as if it had been someone's 'best friend', but it is difficult to know precisely what the concept of 'friend' meant to the people who carried out the burial.[10]

Some dog burials formed part of a sacrificial ritual. The ancient Egyptians, notorious for breeding, killing and mummifying domestic cats by the million, did the same to dogs, though to a lesser degree. At roughly the same time, some 2,500 years ago, the Persians living in today's southern Israel created vast dog cemeteries. Archaeologists have excavated over 1,200 dogs and puppies from one site at Ashkelon, concluding that the majority were not pets but feral street dogs, many of which apparently died from natural causes. No written records indicate the spiritual significance of these interments or why the Persians had stronger feelings for strays than for their own dogs. During the millennium before the birth of Christ, conceptions of dogs evidently varied widely: it is thought that the Hittites, who inhabited what is now eastern Turkey, attached special healing powers to puppies, both living and deliberately sacrificed. Dating to 1,000 years earlier, one graveyard in China contained over four hundred dogs, each beneath a human, indicating that these dogs had been killed when their masters died and interred with them. In the oldest dog cemeteries discovered in the United States, in Tennessee's Green River Valley, some dogs were buried alone, while others were buried alongside people.

The details of the interment sometimes enable us to guess at why a particular dog was buried. At one of the earliest European sites, Skateholm in Sweden, archaeologists discovered fourteen dog graves dating to around 6,000 years ago. In four cases, the dogs were buried alongside people, though most had their own graves, mainly at the edge of the cemetery in an area where the graves of children were also concentrated – as if dogs and children were somehow considered equivalent. At least one dog had received an elaborate burial, its grave strewn with ochre. Alongside it lay grave goods, usually only found in human burials, precious items provided to accompany the animal to the afterlife. In this case these included three knives made from flint and an elaborately decorated hammer made from a red deer's antler (see nearby illustration).

While dog burials continued throughout recorded history, the practice seems to have diminished as societies became more settled and adopted agriculture, with few recorded in Europe since the end of the first millennium. One theory holds that many of these more

A dog's grave at Skateholm.

recent European burials reflect the special relationship between dog and hunter rather than pet and devoted owner.[11]

The majority of dogs given special burials may have been, rather than pets first and foremost, either favoured hunting companions needed by their masters in the afterlife or unowned dogs sacrificed for some spiritual or superstitious purpose. A few examples do point, however, to a primarily affectionate relationship. One of the dogs buried some 7,000 years ago at an ancient cemetery in Anderson, Tennessee, had suffered several injuries during its lifetime, each of which had healed. This dog had grown old enough to suffer from arthritis, and at least for the last few years of its life would have made a poor hunting partner. This suggests that the dog's owner took care of it out of pure affection.

Although women generally played a lesser role in hunting than men, dogs were interred with them too. One such burial from just over 4,000 years ago, found in today's United Arab Emirates, is remarkably reminiscent of the Natufian grave from some 8,000 years before. Dating from roughly the same era but halfway across the world, in Indian Knoll, Kentucky, one graveyard contained six dogs buried with women, six with men, and eight buried alone. Dogs buried with women would most plausibly have been pets.

Other signs suggest that in some cultures dogs were becoming, if not pets as we think of them today, then at least creatures with personalities of their own. Around 3,000 years ago, the Egyptians buried some dogs in a manner indicating that they were treasured more for their companionship than for their practical uses. The hieroglyphs on their gravestones tell us that the Egyptians gave some of their dogs human, rather than distinctively animal, names, echoing the replacement of 'Fido' and 'Buster' with 'Max' and 'Sam' in the West towards the end of the twentieth century.

We can thus trace the affection felt by today's dog owners for their pets back thousands of years in a continuous thread, through our agriculturalist forebears to our more distant hunter-gatherer ancestors. Throughout that time, owners valued domesticated dogs as hunting partners and guards, but the underlying basis for the relationship was the affection dogs showed their masters, demonstrated by their trainability. Dog owners down the ages both recognized and reciprocated that affection: the burials of dogs alongside their doting masters (and mistresses) provide the most durable evidence from the prehistoric period of affectionate relationships that lasted for the whole life of the animal.

WHAT OF CATS? Cats first became domestic, in the sense that they hunted within human settlements, somewhere in the Fertile Crescent about 10,000 years ago, but the first evidence of fully domesticated pet cats appears only about 3,500 years ago, in Egyptian artworks. The ancient Egyptians kept many kinds of exotic animals as pets, including monkeys, cheetahs and small deer, but were nearly obsessed with domestic cats. Their more formal art often depicted cats as the companions of aristocratic women – their husbands preferring to pose with their dogs – but we also have evidence that pet cats became a feature of many, perhaps most, households in all strata of society, since they often feature in sketches done by temple artists for their own amusement. The Greek historian Herodotus reported 2,500 years ago that the Egyptians so venerated their pet cats that when one died from natural causes, the whole family shaved their eyebrows as a mark of respect. The ancient Egyptians undoubtedly also valued cats for their skills at controlling vermin – seemingly finding their ability to deter snakes especially impressive – but prized them equally as pets.

Cats subsequently spread from Egypt around the Mediterranean and, thanks to Phoenician traders, had reached England by 2,300 years ago. People valued them foremost as hunters, however, not as pets. Still, it is difficult to imagine that the off-duty cat snoozing by the fireside did not engender affection wherever and whenever it found itself.[12]

As CIVILIZATION PROCEEDED and small-scale hunter-gatherer societies gave way to urban elites and subservient rural populations, pet keeping entered a completely new phase. In the generally egalitarian communities of the Paleolithic everyone could keep animals as companions, whereas in the highly stratified societies of the Egyptian, Greek and Roman empires, and right up until the twentieth century, the poorest had little opportunity to acquire pets for their own sake. That's not to say that they didn't feel affection for dogs and cats, but those animals had to earn their keep. The surviving evidence generally suggests that from the classical period (fifth and fourth centuries BC) until the end of the nineteenth century, pets played a part in the lives of the wealthiest members of society. As the less well-off inevitably left fewer traces of their lives, we can only guess at how they interacted with their animals; no doubt they had less time and fewer resources to devote to them. Not until the nineteenth century, with the rise of the middle classes, did the keeping of pets for their own sake became widespread once more.

The visual arts of the classical period reveal the elevated status of domestic pets, especially dogs. Greek tombs depict dogs and occasionally cats gazing adoringly at their masters, while children's tombs sometimes include representations of birds. Greek art of the period includes representations of cats – for example, a kitten sitting on a child's shoulder – that clearly indicate their occasional status as pets. Ancient Romans who bred toy dogs can only have intended them as companions (the well-known Pompeian mosaic of a chained dog, inscribed '*cave canem*', shows that many other dogs served primarily as guards). A carving on a Roman tomb depicts a fashionable lady with what looks like a lapdog peering out from under her armpit (see nearby illustration). Dogs also appear on children's tombs, some quietly curled up, others seeming to invite the child to join in a game. On the opposite side of the world, in China, the aristocracy kept lapdogs that bear a striking resemblance to today's Pekinese.

Tempting as it may be to assume that all these dogs were simply pets, they may also have carried a symbolic significance for their owners. The ancient world already regarded dogs as symbols of fidelity – after some 10,000 years of domestication, they probably felt the degree of attachment to their owners that we see today. Also,

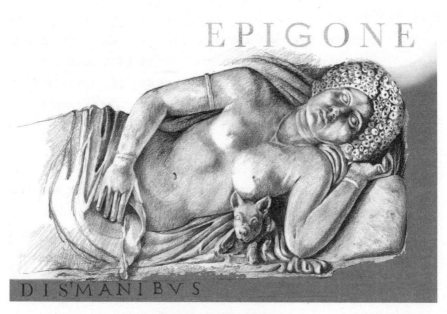

Tomb of Ulpia Epigone.

the ancient Greeks and Romans believed that dogs (and many other animals) had links to the underworld, perhaps explaining why they appear so often on tombs.

The switch from hunting and gathering and nomadism to settled agriculture and animal husbandry seems to have brought about a profound change in people's treatment of animals. Certainly the major monotheistic religions of the Fertile Crescent – Judaism, Christianity and Islam – all emphasize dominion over animals: 'God blessed [the humans] and said to them, "Be fruitful and increase in number; fill the earth and subdue it. Rule over the fish of the sea and the birds of the air and over every living creature that moves on the ground."'[13] The Christian Church generally looked askance at any display of affection towards animals in general and pets in particular: in the thirteenth century the Franciscans (founded by the animal-loving St Francis of Assisi) were taken to task by the authorities for their fondness for dogs, cats and small birds. The reviling of cats as potential agents of Satan apparently stemmed from the pagan worship of cat gods and goddesses in rural areas of Europe. While generally regarding dogs as unclean, Islam viewed cats rather more positively, with the earliest cat sanctuary reputedly founded in Cairo in 1280. Only Buddhism consistently emphasized respect for non-human animals, embedded in its concept of reincarnation.[14] Whether this institutionalization of monotheistic attitudes brought about a change in attitudes towards animals or merely legitimated a new necessity for productivity in societies that now relied heavily on animals for both food and transport, the result was a great deal of what we today regard as cruelty.

During the early Middle Ages (the fifth to tenth centuries CE) attitudes towards domestic animals were largely utilitarian, at least in western Europe. The ninth-century poem 'The Scholar and His Cat, Pangur Bán', written by an Irish monk, compares the poet's struggle to find insight with his cat's mousing:

> 'Gainst the wall he sets his eye, full and fierce and sharp and sly
> 'Gainst the wall of knowledge I, all my little wisdom try[15]

Notions of 'indoor' and 'outdoor' animals
have changed since medieval times.

The poem's eight stanzas mention no affection he might have felt for the cat, only admiration for his prowess as a hunter. Tenth-century Welsh statutes valued a female cat at four pence, but only for her mouse-catching and breeding abilities – not for her readiness to purr on someone's lap. An untrained dog went for four pence, but the price doubled after its training, implying its value lay in the tasks it could perform.

The period between the eleventh and fourteenth centuries saw little change in attitudes towards animals in general, and pet keeping remained limited to those who could afford it – and could afford to ignore the disapproval of the Church. Fashionable ladies continued to keep lapdogs. Any praise of working dogs for their faithfulness may simply have reflected how easy this made them to train. Farmers often gave names to individual animals, sheep for example, but not necessarily out of any particular affection. Cats became identified with their agricultural function, not as household pets, as in this passage from Daniel of Beccles's book of advice for aspiring noblemen, the *Urbanus* (see illustration on p. 45): 'Let not a brute beast be stabled in the hall, let not a pig or a cat be seen in it; the animals which can be seen in it are the charger and the palfrey, hounds entered to hare, mastiff pups, hawks, sparrow-hawks, falcons, and merlins.' Monks especially valued cats for their fur, which, being cheaper than fox fur, did not violate their vow of poverty: fourteenth-century East Anglians fixed the price of 1,000 cat skins at just four pence.

During the sixteenth century pet keeping continued to thrive among the well-to-do, especially women, and the word 'pet', in the sense of animal companion, first appeared in the English language. John Caius's 1576 book *Of Englishe Dogges* divides the species into two kinds: peasant dogs, or 'curs', and 'noble' dogs, which included dogs for hunting and retrievers for hawking (both largely the province of men) and lapdogs, which noblewomen continued to favour (see nearby illustration). Caius's description leaves little doubt that the latter were pets in the modern sense of the word: 'These puppies, the smaller they be, the more pleasure they provoke, as more meet playfellows for mincing mistresses to keep in their bosoms, to keep company withal in their chambers, to succor with sleep in bed, and nourish with meat at board, to lay in their laps and lick their lips as they ride in wagons.'

Cruelty, even to dogs, was widespread during the Renaissance. The term 'hangdog' derives from the habit of killing old or injured dogs by hanging. People subjected cats to all kinds of treatment that we would condemn today. By no means the most extreme was a form of entertainment known as '*Katzenmusik*', which consisted of tying strings of bells to several cats, cramming them into a sack, and then letting them out in an arena to fight, their growls and howls accompanied by the jangling of the bells. Going to bear- and bull-baiting events that used dogs was a perfectly acceptable alternative to attending the theatre for a performance of William Shakespeare's latest play.

Not until the seventeenth century did pet keeping become widespread. Before that, houses in towns, like those in the countryside, had been full of animals – pigs and poultry as well as dogs and cats –

Kornelis Visscher's sketch of a 'butterfly dog' wearing a collar with bells attached, c. 1650.

blurring the distinction between companionship and cohabitation. Yet Church strictures against feeling affection for animals remained uppermost in some people's minds: in 1590, even as she lay dying, Katherine Stubbes beat her favourite puppy, believing she and her husband had 'offended God grievously in receiving many a time this bitch into our bed'.[16]

A gradual change in the perception of animals accompanied progressive urbanization: rather than seeing them as mere machines, as René Descartes and other philosophers suggested in the mid-seventeenth century, the general public widely accepted them as capable of not just receiving but returning affection. Thomas

Bedingfield first proposed the benefits of harnessing this sentiment for practical ends: in *The Art of Riding*, a 1584 translation of an Italian manual, Claudio Corte's *Il Cavalarizzo*, he told horse trainers that ensuring their charges' love for them was more effective for training than the previous harsh methods based on 'mastery'. Monkeys were popular pets, valued for their ability to 'ape' human behaviour. For the first time, cats became popular companions, especially for women, but small dogs were still more prevalent, frequently appearing in portraits. The practice of burying favourite dogs in special cemeteries re-emerged. Whatever the species of animal, the concept of mutual affection came to be widely accepted.

Of course, the increasingly tight bond between some humans and some animals did not put an end to animal cruelty (nor has it still). The burning alive of cats enjoyed wide acceptance up to the seventeenth century, and until 1817, the Festival of the Cats, celebrated to this day in the Belgian city of Ypres, featured the throwing of a bag full of live cats from the top of the church tower (nowadays the bag contains soft toys). In the countryside, attitudes towards dogs could be far from sentimental: in 1698 a Dorset farmer recorded his satisfaction at having extracted eleven pounds of grease after killing and then simmering his elderly dog.[17]

WHEN PET KEEPING for its own sake began to expand during the eighteenth century, the choice of species was if anything greater than it is today. The majority of animal companions were not domesticated species but tamed wild specimens, an unwitting reflection of the habits of hunter-gatherers in distant and still uncharted parts of the world. Pet tortoises, monkeys, otters and squirrels were all readily available for those who could afford them, but perhaps most popular in eighteenth-century London were caged songbirds (canaries and chaffinches were particularly affordable) and talking jackdaws, magpies and parrots. The poet William Cowper (1731–1800) kept three pet hares that he named Puss, Tiney and Bess (only to discover afterward that they were all male). His devotion to them extended to having a snuff box made with an engraved lid depicting the three and listing their names. They were evidently less devoted to him, since Puss in particular made regular escapes requiring forcible retrieval. Cowper later kept a series of three pet dogs: Mungo, The Marquis, and then Beau; his biography describes the latter as follows: 'Whether frisking amid the flags and rushes, or pursuing the swallows when his master walked abroad, or whether licking his hand or nibbling the end of his pen when in his lap at home, Beau ofttimes, like his predecessor, the hare, beguiled Cowper's heart of thoughts that made it ache, and forced him to a smile.'[18]

Dogs kept exclusively as pets were still probably rare, but some owners evidently regarded their working dogs with affection. By the early eighteenth century, a farmer's favourite hound might live indoors: 'Caress'd and lov'd by every soul, he ranged the house without control.' By the end of the nineteenth century cats had become popular pets, made fashionable in the United Kingdom by Queen Victoria. Meanwhile, across the Atlantic, Mark Twain wrote, without irony, in an essay published after his death in 1910, 'When a man loves cats, I am his friend and comrade, without further introduction.'[19]

OVER THE BROAD SWEEP of prehistory and then history, pet keeping went through two distinct stages. Even prehistorically, mankind had a far more complex perception of animals than simply that of the predator for his prey, the hunter for the hunted. Though always a valued source of protein, animals at some point – perhaps as sophisticated consciousness first evolved in the hominid brain – took on other significances, if the customs of surviving Paleolithic peoples are anything to go by. Humans chose some animals to share their living spaces, even integrating them intimately into the family. The widespread breast-feeding of young mammals, shocking to modern sensibilities, might superficially suggest a bond very different from that between today's pet and its owner. However, it probably arose as a straightforward nutritional necessity – as the only way to raise baby mammals captured before weaning. The practice does, nevertheless, point to a powerful and apparently near-universal instinct among hunter-gatherers to extend their most intimate caring to, and expend essential resources on, young animals.

This first stage gradually phased out as hunter-gatherer groups gave way to societies stratified into rulers and subjects. In the second stage, pet keeping became the privilege of those with money and influence. A common thread runs through both: a difference between the sexes in their fundamental attitudes towards animals. In small-scale societies, women and children often took the most care of captured wild animals. In the Middle Ages, while aristocratic men valued hounds and hawks for their utility and the prestige they conferred, their ladies demonstrated affection for specially bred small dogs.

Thus, while frowned on for much of recorded history by the (almost entirely male) authorities, pet keeping continued in everyday life, sustained largely by women and primarily, but probably not exclusively, by the well-to-do. Societies occasionally attempted to suppress this: in medieval Germany, thousands of women stood accused of witchcraft based on their affection for their cats. Even in the late seventeenth century, those condemned during the Salem witch trials included two dogs 'possessed by the Devil'.

By the eighteenth century attitudes had begun to change, paving the way both for a much more humane approach to domestic animals in general and for the third stage of pet keeping: its universal acceptance in the West.

CHAPTER 2

. . . and Modern

MODERN PET KEEPING seems, on the surface at least, nothing like the practice of our hunter-gatherer ancestors. Our predilection for sentimentality may have found its ultimate outlet on the Internet, but even today's print media serve as vehicles. Under the headline 'Puppies, Kittens and Bunnies Are Born', a leading article in the 4 April 2015 edition of *The Times*, proclaimed the inauguration of the paper's newest – and perhaps most tongue-in-cheek – section, 'Pet Announcements'. 'We are pioneering anthropomorphic journalism,' the article declared. 'Humans would be immeasurably poorer without pets' – presumably in the emotional sense, given the burgeoning list of opportunities available for pet owners to part with their cash. 'They comfort us, amuse us, play with our children, return our affection . . . and assist us when we are lonely, frail or disabled.' The pages themselves interspersed the traditional births, deaths, birthdays, and even anniversaries (on that particular day, the third wedding anniversary of two rabbits) with more animal-specific sections titled 'Seeking a New Home', 'Lost', and 'Quirky Habits'.

The belief that pets are not simply companions but have the power to educate us, console us, and even to prevent us from becoming sick has become all-pervasive. These are superficially modern developments, but our prehistoric ancestors may have held similar beliefs. Why else, for example, would they sometimes bury their pets with such ceremony? Our attitudes towards pets may seem unprecedented in their specifics, but who is to say that they do not arise from much more fundamental, almost instinctive ways that we have always re-acted to animals?

Our present-day attitudes towards pets stem from the separation between home and workplace that occurred during the Industrial Revolution. In traditional rural societies, and even in many towns, humans and domestic animals shared a living space. Cities teemed with animals: horses for transport, chickens for eggs, cows for milk-ing. Until the nineteenth century, butchers slaughtered animals at the rear of their shops, above which they often lived. Purpose-built out-of-town abattoirs did not emerge until people began to complain about the cattle, sheep and pigs crowding the streets, not to mention the questionable hygiene of the slaughtering arrangements.

During the Victorian era, animals and humans increasingly oc-cupied separate spaces. With the advent of the railways, cows and chickens no longer needed to live among humans: trains could trans-port their milk and eggs from the countryside into the city. The housing of food animals in their own accommodation honed the distinction between 'outdoor' and 'household' animals. For most people, however, this distinction only applied to individual crea-tures, not to whole species. While people might lavish dogs living inside the home with affection, they regarded stray dogs not simply as nuisances but as savage wild animals, fit only for elimination. The French and other Europeans still customarily killed stray canines by hanging; Americans preferred lassoing and strangulation. Neverthe-less, the distinction began to sharpen between animals kept primarily for their companionship and those that were merely useful.

Changing attitudes towards domestic animals led to the founda-tion of animal-protection societies. Reformers established the first, the Society for the Prevention of Cruelty to Animals (SPCA), in London in 1824 (Queen Victoria granted its 'Royal' prefix sixteen years later), primarily to protect the horses and donkeys working on the streets of London, and eventually domestic pets as well. Simi-lar organizations followed in all the major European capitals. By the late 1860s numerous SPCAs had appeared in US cities and also in

Montreal, the first such organization in Canada. These attitudes favouring the humane treatment of animals would pave the way for an explosion in family pet keeping.

By the end of the nineteenth century, the keeping of pet dogs had assumed its current form. Dogs still played their traditional roles, providing social status for men and company for women. But they also became emblems of taste with the advent of distinct pedigreed breeds, which had become common by the early twentieth century. Parisians (naturally) not only paraded their dogs as fashion accessories but dressed them in haute couture: 'pretty embroidered coats, silk jackets, warm outfits for the winter, light ones for the summer,' remarked one commentator.[1]

At around the same time, a new type of sentimental literature emerged throughout the West, depicting dogs as selfless heroes, exemplars of fidelity and loyalty. The legend of Greyfriars Bobby was one such (see Chapter 6); another, by the Scottish SPCA's founder, tells of Buck, a Skye terrier living in Paris, who attended the funeral of each member of his family, one after the other, and then finally 'lay down in an agony of despair, and with a mournful cry, which spoke of the depth of his emotion, expired'.[2]

CATS DID NOT become broadly accepted as pets until the early twentieth century. Some fifty years earlier, popular opinion still deemed them too 'bohemian' and insufficiently loyal to rival the ever-faithful and self-sacrificing dog. A pamphlet produced by the RSPCA warned potential cat owners, 'Even in her best mood she is not always to be depended on.'[3] It didn't help that neutering was not widely available for cats – nor would it be until the 1930s – and the cat's noisy and overt sexual behaviour jarred Victorian sensibilities.

Cats also had their champions, however – notably illustrator Harrison Weir (see nearby illustration), who organized the first cat show in 1872 at London's Crystal Palace. The promotion of cat breeds that differed distinctly from the common moggie/alley cat helped improve the whole species' image.

An illustration by Harrison Weir, founder of the modern cat
show, for the moralistic children's story 'Three Little Kittens'.

By the late nineteenth century, the habit of keeping cats as pets
had become fairly widespread in the West. But still encouraged to
hunt, domestic felines continued to have one paw in the wild. Most
owners, indeed, did not keep them as house cats but allowed them to
roam as they wished, even deliberately shutting them out overnight.
Then in the 1950s the cat's hunting role began to wane with the development of effective chemical rodenticides. Many cats continued
to hunt, however, of necessity; until the 1970s much commercially
available cat food was deficient in one or more key nutrients. Not
until the early 1980s did nutritionists fully appreciate cats' metabolic

quirks, which extend to the whole cat family, and design today's nutritionally complete foods. The more squeamish among us probably always preferred not to think about how our cats sustained themselves, but over the past three decades reactions towards the dismembered rodents that cats supposedly leave as 'presents' have changed from general admiration to horror and disgust. In the popular imagination, assisted by their portrayal on the Internet, cats have evolved from silent killers into sloppy cuties – and we now expect them to behave like devoted pets 24/7.

FOR THE LAST 150 years, dogs and cats have been far and away the most popular pets in the West, but of course people keep other kinds of pets as well. The nineteenth century saw a proliferation of captive wild animal pets in middle-class households. Owners prized native species less than those from exotic locations, perhaps as emblems of European nations' dominion over their far-flung empires. Decorative aquaria of tropical fish became increasingly popular, as did cage birds such as parrots, budgerigars and canaries. Of course, many people today own exotic birds, fish, snakes, turtles and tortoises, but these remain minority interests, accounting for 1 per cent of households in both the United Kingdom and the United States. Likewise 'pocket pets' – mice, gerbils, hamsters, and the like – are popular with parents of young children but rarely become a lifetime interest; thus they account for only about one in fifty pet-owning families. Because they have to be kept caged for much of the time, they rarely generate the same level of sustained affection as the much more interactive dog or cat.

Rabbits are emerging from their original status as caged animals. Until the 1980s, pet rabbits lived in small hutches, often outdoors, and served as cheap 'starter pets' for children (people felt little concern about whether such a lifestyle was ideal for a free-ranging, gregarious species). More recently, with the introduction of the toilet-trained house rabbit, they have become an increasingly popular, if still minority-interest, house pet, receiving the same veterinary care, toys, high-quality food and dedicated space as many indoor-only cats.

Even more remarkable is the rise of the leisure horse. Over the course of its history, the horse has undergone multiple transformations. In prehistoric times, horses primarily provided milk and meat. By 4,500 years ago, they had become draft animals, as shown by the earliest chariot burials, which typically contain a warrior, his chariot and his favourite horse. The twentieth century witnessed the completion of their transformation into beloved companions.

The invention of the internal combustion engine should have consigned the domestic horse to oblivion. Yet the number of horses in the United States has increased at least threefold since 1950 and in both the United States and United Kingdom has risen by about one-third over its previous peak in 1900. In 2005, the US leisure-horse industry was worth an estimated $80 billion to the economy. (Given that horse ownership runs at about one-tenth that for dogs, this sum indicates just how expensive it can be.) To quote a recent survey of US horse owners, 'If the ownership of a dog or cat costs roughly as much as the purchase of a slightly used Buick, the cost of horse ownership over a 25–30 year lifetime could conceivably buy a new Maserati.' Perhaps not surprisingly, the time and money that horse owners must devote to their animals can create conflict within their (human) families.[4]

While often deeply beloved by their owners, horses are not, biologically, domestic pets. Not only are they are obviously too large to live in modern houses, but they form much stronger social bonds with other horses than with people. Unlike dogs and cats (and probably rabbits), they cannot be socialized to humans at a young age. Rather, they must be 'broken' – gradually habituated to the proximity of people – before they can enter any kind of individual relationship with their owners. Despite some parallels to the relationships people have with dogs and cats, significant differences lead most authorities to class horses as companion animals but not as pets.[5]

THE CURRENT EXPLOSION in pet ownership in the West dates back to the economic boom that followed World War II and the exodus from cities to the new, pet-friendlier suburbs. In the twenty-first century, ownership of cats and dogs far surpasses that of any other kind of pet. In the United States, more than one-third of households have dogs, while slightly fewer have cats. The United Kingdom, despite its reputation as a nation of pet lovers, has somewhat lower numbers: one-quarter of households own dogs and one-sixth own cats. Cats are more widespread in continental Europe, occupying around one-quarter of households, and outnumber dogs, found in roughly one home in five. The available physical space is an important factor: dogs generally need more than cats, and rural families thus tend to favour dogs, while city dwellers tend to prefer cats (seemingly unaware of the difficulties many cats experience living in high-density cities, as highlighted in a recent series of BBC Horizon documentary, *Cat Watch 2014*). In Japan, dogs are uncommon in small apartments but share roughly a third of households with access to more than one hundred square metres of space. In a survey conducted in Australia, over half of those who didn't currently own a pet would have liked to, if circumstances permitted.[6] Many people do own cats or dogs, but practicality, as much as proclivity, affects their choice of which: the recent increase in the popularity of cats parallels the ever-growing proportion of people who live in towns and cities, where dogs require more effort than they would in the countryside.

Practical considerations aside, fashion plays a major role in the type of pet people choose. Every few years a craze for some new type of animal helps its owner stand out in the crowd. Miniature pigs are just one recent example. It's unclear whether fashion, rather than lifestyle, influences the choice of cat or dog – but there have certainly been plenty of fads for specific breeds of dog, especially in the United States. A hit movie featuring a particular breed can catapult it into popularity. Old English sheepdogs became favourites in the 1960s following Disney's 1959 *The Shaggy Dog*. The original 1950s Lassie movies produced a surge in ownership of rough collies, but the breed's popularity has declined gradually, possibly because it needs a great deal of exercise and therefore doesn't suit many contemporary lifestyles. (The same factors may also account for the rapid rise and fall of the Dalmatian following the 1985 remake of *101 Dalmatians*.) Registrations of Labrador retrievers increased following Disney's 1963 *The Incredible Journey* – and continued to do so for the next two decades, possibly because the Labrador doesn't demand too much in the way of exercise. By contrast, numbers for the other breed featured in the same film – the bulldog – barely changed. The Chihuahua alone seems to buck this trend: even though *Beverley Hills Chihuahua* (2008) performed well at the box office, it failed to halt the breed's decline from eleventh position in 2006 to twelfth in 2009 to twenty-fourth in 2014. A slight increase in its numbers in the late 1990s may have had something to do with the popularity of 'Gidget', a Chi-huahua (played at different times by two or perhaps more 'actresses') featured in television advertisements for Taco Bell. Quite possibly, with the proliferation of new media forms, individual movies no lon-ger have the impact they once did on the lifestyle choices of viewers.[7]

More fundamental cultural changes have also affected the popular-ity of different types of pets. Westernization has transformed the cul-ture of dog ownership in Japan. Western breeds such as toy poodles, dachshunds, Pomeranians and Yorkshire terriers – and for the lucky few with more space, golden and Labrador retrievers – have largely displaced traditional Japanese breeds, including the Shiba Inu, the Kishu Inu and the Akita – *inu* is simply the Japanese word for 'dog'. The Japanese have come to treat their dogs differently as well, having formerly kept them mainly for guarding and hunting, and so chaining them up in the garden when they were not working. The concept of the dog as family pet is relatively new to Japan, which probably imported it from the United States along with American-bred dogs. Part of the same trend is the increasing popularity of dog training; the Japanese previously only trained hunting dogs.

In certain ways, however, Japan remains its own culture when it comes to dogs. Even today, Japanese people of all ages tend to disapprove of euthanasia, whereas vets in the West routinely put down suffering dogs that have little hope of recovery. The West-ern approach is probably a residue of the Cartesian/Christian belief

Traditional Japanese dog breeds are on the decline.

in dominion over animals, whereas the Buddhist–Shintoist tradition conceives of humans and animals alike as part of nature, which, in some interpretations at least, entails the right to suffer. This fundamental cultural difference has also led to difficulties in winning acceptance of guide dogs for blind people in Japan. In the United Kingdom and United States, these dogs enjoy the status of 'partners', taking responsibility for their blind guardians' well-being, albeit under a necessary degree of control. However the Confucian traditions in Japan create the perception that guide dogs are slaves: their duty to go everywhere their blind human goes deprives them of their natural right to autonomy.[8]

So, while some aspects of pet keeping seem to travel well from one culture to another (usually as part of so-called coca-colonization, the global spread of American customs and values), others do not, rendering the lives of otherwise identical pet dogs – and working dogs – somewhat different, depending on where they happen to live. In general, however, in the post-industrialized world, the bond between owner and pet is growing ever stronger and ever more personal.

IN THE TWENTY-FIRST CENTURY, household pets carry more emotional significance for their owners than at any time in the past. According to surveys conducted by the American Animal Hospital Association (AAHA), while many respondents (47 per cent) would take a human as their sole companion on a desert island, more – 50 per cent – would take a dog or cat. In answer to the question 'Who listens to you best?' 30 per cent specified their spouse or significant other; half as many again (45 per cent) specified their pet. Four out of five said they thought about their pet several times a day while away from home. Over half claimed that they would 'very likely' risk their lives for their pets. Over two-thirds admitted to breaking household rules, such as allowing a pet on the bed, when their partners were away.[9]

Given the intensity of the bond between modern humans and their pets, it is hardly surprising that spending on them has skyrocketed – and increases every year, regardless of the state of the economy. In the 2014 AAHA survey, three out of five owners reported spending more on their pets than they had three years before, while 41 per cent reported spending more than $500 on veterinary bills for their pets in the previous twelve months. Only a quarter of respondents visited their own physician more often than they took their pet to the vet.

More than spending on pet care has risen. The cost of the pets themselves can be astronomical. Certain breeds have become luxury items and status symbols – little different from a Mont Blanc pen, a Porsche sports car, or a Patek Philippe watch. Down the centuries, until the passage of legislation to protect their welfare, big cats such as lions and tigers were the 'status cats' of choice among the nouveau riche (many of whom, having no clue how to care for them or indeed how dangerous they can be, subsequently abandoned them). Nowadays, 'designer' dogs and cats seem to have filled this niche. A perfect Ashera cat – a hybrid of the domestic cat and the wild serval, with 'snow leopard' markings on its coat – will set its owner back a cool $125,000. In China, rising personal wealth, coupled with demand for traditional but hard-to-find breeds, such as the Tibetan mastiff, has caused prices to skyrocket to over $1 million. In the United States, a pure-white Samoyed puppy may cost well over $10,000; in the United Kingdom in 2014, a rare 'lilac-coated' British bulldog puppy valued at £16,000 was stolen from its breeder's home.

While paying tens of thousands of dollars for a pet may seem excessive, some owners go even further, paying hundreds of thousands or even millions to clone their favourite animals. Ever since the technology became available some fifteen years ago, a small number of devoted owners have fantasized about replacing a treasured pet with an exact replica. In 2000, Texas billionaire John Sperling founded the company Genetic Savings & Clone specifically to reproduce his beloved border collie/husky mix Missy. The 'Missiplicity Project' failed: as it turned out, cats were easier to clone than dogs, but there were few takers for the handful of cats so produced and priced at $50,000 apiece. Advances in the technology (somatic cell nuclear transfer) at Seoul National University in South Korea, however, eventually led to the production in 2007 of not one but four 'Missies'. Two years before, the same team had created Snuppy, the world's first cloned dog, an Afghan hound gestated in a yellow Labrador retriever. The first dog clone sold commercially was Lancelot Encore, a male yellow Lab commissioned by a Florida couple for $155,000: the dog went on to score another first by siring a litter of puppies of his own in 2012. Other clients of the Seoul laboratory have included a vet from Louisiana who had two clones, Ken and Henry, produced from his 'Catahoula leopard dog mix', named Melvin, originally bought for $50. The same company, Sooam Biotech, attempted to stir up interest in the United Kingdom by running a competition with a cloning as the prize: the winner was Winnie, an elderly dachshund who, owner Rebecca Smith claimed, had helped her overcome bulimia as a teenager.[10]

While dog cloning is now possible, some owners may find the results disappointing. The human attachment to animals rarely rests

Cloned dogs.

on aesthetics alone. People who clone their dogs are likely hoping not simply for an exact physical replica of their beloved pet but also for a replacement with the very same personality and habits. Unfortunately both nature and nurture, genes and environment, shape a puppy's personality as it matures.

PEOPLE WHO OVERSPEND on their pets can tarnish the image of pet keeping in general; so too can animal hoarders. Those who collect animals – usually cats or dogs – in numbers that compromise their own and the animals' heath may appear to have arrived at the unwholesome end of a spectrum of emotional involvement that ultimately includes all pet owners. Alternately, their obsession with animals may point to an underlying mental disorder. In either case, their behaviour sheds little light on normal pet keeping.

There may be as many as 3,000 full-blown pet hoarders in the United States alone, but the high rate of reoffending, which may reach 60 per cent, makes estimating their numbers complicated. Once acquired, the habit of collecting animals seems hard to break, and some perpetrators go to extremes to avoid forceful separation from their furry companions. In one notorious case, Vikki Kittles (aka Susan Dietrich, Rene Depenbrock or Lynn Zellan) travelled the length and breadth of the United States in a bus crammed with over a hundred dogs, before finally being arrested in Oregon in 1993. Undeterred by her jail term for animal cruelty, she began collecting dogs again in 1996. She switched to cats in 2000 and was prosecuted in Wyoming in 2002, only to resurface – with a maggot-infested dog in tow – in Colorado in 2006; in 2012 a Wyoming court convicted her of cruelty to eighteen cats.[11]

Animal hoarders tend to fit the 'crazy cat lady' stereotype. Most are female, single and over sixty years old, and most collect cats (although men tend to collect dogs). Hoarders consider themselves supreme animal lovers, believing that no one else can give their animals the care they do. Many are also convinced that they have a unique ability to understand and communicate with their animals.

Might any of these behaviours resemble regular pet keeping? The hoarders' warped worldview is of little significance for understanding typical companion-animal ownership. Their behaviour may have much in common with compulsive hoarding in general, the difference being that a collection of old newspapers is a fire hazard but can otherwise be ignored, whereas a collection of animals requires constant care and expenditure. These hoarders may simply find the urge to collect too strong to resist. Again, since a condition akin to obsessive–compulsive disorder seems to drive these people, their choice to collect animals probably tells us little about genuine pet keeping.

The scientific literature on animal hoarders suggests that they differ greatly from regular pet owners. A significant percentage of them lack attachments to other people, usually because they have survived chaotic or traumatic childhoods. Many show signs of mental illness, although no one particular type predominates. They transfer all their primary affections to animals: a fifty-year-old woman said of one of her male cats, 'I think that Freddie thinks I am actually his wife in a way because he is so close to me.' They have overpoweringly anthropomorphic relationships with their animals, considering them not simply part of their family, as many pet owners do, but to actually *be* their family. They deem their animals at least as intelligent as humans and believe they can communicate with them on an equal basis.[12]

In some of its manifestations, therefore, animal hoarding may appear to be an extreme form of pet keeping, but it is most likely to stem from mental illness or an extreme personality disorder in which the focus on cats or dogs is almost coincidental. The term 'hoarder' describes a particular situation for the animals concerned, not the state of mind of the person (ir)responsible.

Hoarders seem to differ from regular pet owners in the extreme emotional gratification they obtain from their animals and the unshakeable conviction that only they have sufficient insight into their animals' needs to care for them adequately (hence their unwillingness to surrender them to those in a better position to care for them). In moderation, both of these qualities form part of normal pet ownership; in hoarders they often interact with other, unrelated factors, such as a history of abuse, to become pathological.

SOCIETY'S RECENT FOCUS on another extreme manifestation of pet keeping, so-called dangerous dogs, also threatens to undermine public opinion of companion dogs in general. Some owners regard their animals, dogs especially, as extensions of themselves and choose a breed to enhance their self-image. This is not necessarily harmful in itself. Ownership of a golden retriever, perhaps the archetypal 'family dog', projects 'family values'. Rottweilers and Dobermanns project machismo; Persian cats and Pomeranians, the opposite. Such projections are generally healthy in moderation. In the extreme they can lead to practices unacceptable by today's standards. Consider the owner who, in his desire to dominate, prefers the so-called fighting breed, such as a pit bull, and trains it to be aggressive.

Staged contests between animals were once considered a perfectly acceptable form of entertainment. Nowadays, they violate modern conceptions of animal welfare. They also create dogs that do not conform to the modern expectation that canine companions should behave 'well' (non-aggressively) in public – hence the public outcry when a dog so much as scares someone.

It is not easy to disentangle the truth about pit bulls and other fighting breeds from the invective that surrounds them, but one study of gang youths in South Wales showed that overt cruelty to animals and strong attachment to individual dogs can, somewhat paradoxically, exist side by side. These youths usually selected bull-type dogs – mostly Staffordshire bull terriers – and used them in encounters with rival gangs: 'for a laugh. . . . [T]here's dog fighting in the park, it just happens. . . . [S]ome of the boys train them for fights. . . . [Y]ou put them into a cupboard and give them only red meat so they get the taste of blood. Then in a week or two after being in a dark room, it comes out wide-eyed and they fight it.' The dogs were sometimes starved prior to fights, and occasionally electrocuted in the belief this would enhance their aggression.[13]

All of these practices fit the stereotype, portrayed by the media and authorities alike, of 'feral' youths and their 'weapon dogs'. Yet the gang members themselves stated that they kept dogs not for competition but for companionship ('she's my best friend. I've had a hard time and she's the one I want to spend time with, she's always with me') and as a common interest with other gang members ('we just walk, play, hang out, walk to the park. . . . [They are] part of the group, like, it feels good'). The study does not make clear how the gang members reconciled the affection they undoubtedly felt for their dogs with the suffering they evidently knew they were inflicting on them. It does show, however, that relationships with dogs, at least, can be much more multifaceted than the simple contrast between their being 'one of the family' or 'a weapon'.

WHILE EXTREME MANIFESTATIONS of pet ownership – be it participation in illegal dog fighting, hoarding of a houseful of starving cats, or purchase of a million-dollar dog – often receive a great deal of media attention, normal, everyday pet ownership is so ubiquitous as to scarcely merit comment. And yet the practice is indeed remarkable, given that the drawbacks of sharing a home with an animal are self-evident, while the benefits are, on the surface, less tangible. So why do people take on pets in the first place?

We must first dispel the myth that people only keep pets because they're lonely. If that were the case, then people living on their own would own most pets. In most countries, though, dog ownership is most prevalent in large households, especially among families with children living at home. Of course, there are practical reasons for this. Some people living alone might feel they cannot have a pet because they won't be home much of the time. The elderly may doubt whether they will be able to meet a cat or dog's needs for its entire lifespan – a consideration borne out by one survey of UK owners, in which cat ownership didn't start to decline until the respondents were about sixty-five, but dog ownership decreased from about fifty-five years old, perhaps because respondents became concerned about their ability to walk a dog every day for the next ten to fifteen years of their lives. In Japan, less than one man in ten over sixty-five owns a dog, even though more than 50 per cent of this cohort have owned a dog at some point during their lives.[14]

Obviously all sorts of practical issues inform people's decisions about whether to keep a pet and which kind. The most important driver of the desire for a pet seems to be whether the individual grew up with animals. The conception of what constitutes a 'household' begins in childhood; once acquired, it sticks. Unsurprisingly, then, people who grew up with dogs tend to have dogs later in life; those who grew up with cats tend to keep cats; those with close experience of both aren't fussy but definitely like to have animals around.[15]

Within pet-owning families, the adults look after most dogs and cats, which may nominally belong to one of the children. However, the small pets that children more typically say are 'theirs' (even when a parent actually does most of the associated chores) are more popular with girls than boys, and with the youngest sibling, as if in compensation for the lack of a younger brother or sister. In this era of fractured families, step-parents may also be more likely than biological parents to obtain pets for the children who live with them, again possibly as a form of compensation.

As adults, we make all sorts of decisions – whether to adopt pets, how many children to have, what church to attend – that have some basis in our early family experience. But could other factors come into play? Might the desire – or lack thereof – to own a pet be genetically heritable? That is, could affection for pets be encoded in our DNA and passed down from parent to child? There is some reason to think this is possible. Religion provides a much better-studied example: culture determines one's choice of faith, but religiosity – the strength of one's belief – is quite strongly genetically inherited. If pet keeping is like religion (and I'm only suggesting a useful analogy, not that pet keeping is a kind of faith), then we might predict that affection for animals has a genetic component.

Psychologists traditionally use studies of twins to determine whether DNA influences behaviour patterns. Twins raised by the same parents generally have the same life experiences until they leave home, but they differ in their genetic similarity to each other depending upon whether they are identical (100 per cent similar) or fraternal (no more similar genetically than any pair of siblings, at 50 per cent). Using complicated mathematics researchers then disentangle the effects of genes, shared environment (generally during childhood) and unique experiences (occurring mostly after the twins studied have left home and lived apart).

One of the earliest and most famous twin studies featured two dogs with the same name. Twins Jim Springer and Jim Lewis had been separated at birth and did not meet until they were thirty-nine. Such rare events are like gold to researchers, enabling them to distinguish between genetic and experiential effects. Both men had pet dogs, which both, astonishingly, had named 'Toy'. The latter was presumably a coincidence, but the fact that both had dogs fits with the idea that pet keeping runs in families.

To date, only one twin study has examined attitudes towards pets, and somewhat obliquely at that, but it yielded unexpected results. The subjects were 500 pairs of twins, some fraternal, some identical. All were male, and most were US citizens of Caucasian extraction between the ages of fifty and sixty; all had grown up with their twin (some twin studies, though not this one, focus on twins who've been adopted separately). The study simply asked whether participants had played with a pet animal during the previous month.

The conclusions were astonishing. Childhood experiences seemed to have little influence: the twins hadn't acquired their desire to interact with animals from their parents' habits. They had, however, inherited their parents' genetic predisposition to interact with animals: the answers given by the identical twins (who are genetically identical) were more closely aligned than those of the fraternal twins (who share only 50 per cent of their genes). Nevertheless, even the identical twins did not always give the same response as each other. Each man's individual experiences, presumably gained mostly during the thirty-five or so years that he and his twin had lived apart from one other, were roughly twice as influential as his genes.[16]

So, growing up around pets doesn't seem to be the crucial factor in whether a man goes on to like pets in middle age. The two appear to be linked, certainly, but the link seems to reside in the genes, not in the childhood home environment. Parents who are genetically well disposed towards animals will tend to have pets, but even if they don't – perhaps constrained by place of residence or economic factors – their children will inherit those genes and go on to like pets when they grow up. Adults without the putative pro-animal genes are unlikely to feel positively towards animals even if they have been raised in a house full of pets. Yet, although living with pets in childhood doesn't seem to have a lasting influence, experience of animals during adulthood does: people who take on a first pet tend to want to repeat the experience.

These conclusions must remain tentative. Experience of pets in childhood may have some influence, but we lack data showing how strong the influence of childhood family environment might be on people who have recently left home and are acquiring their first pet. Maybe parental influence takes thirty years to disappear, by which time most of those who adopted a pet on first setting up house – and then kept it throughout its whole lifespan – would have had more than one opportunity to join the ranks of the disenchanted.

We can thus propose an explanation for the 'family effect', even though it relies heavily on this one study. The desire to acquire a pet does seem to have some genetic basis and may initially find itself most easily expressed in families where both parents like pets enough to have acquired one (or more) while their children were growing up. This acquired habit may persist into young adulthood, but by middle-age the non-genetic part of the 'family effect' has probably disappeared, replaced by the impact of more recent encounters with animals. And we have no information yet on whether the same pattern is true of women – although it might well be, given that genetics generally affect personality traits in men and women equally, with genes having around 40 per cent influence and experience 60 per cent.

PEOPLE WHO GROW UP around pets aren't just more likely to own pets later in life; they are more likely to be concerned about animal welfare in general. They are also more likely to have an interest in wildlife and animal conservation, and to be vegetarians. This effect may be independent of the culture a person grows up in. When Dr Ayaka Miura and I studied the effects of childhood experience of pets on the attitudes of young adults in both Japan and the United Kingdom, we found some remarkable similarities. Even though fewer Japanese had lived with pets as children, those who had were much more positive about pet keeping. In both countries, the more pets people had lived with in childhood, the more likely they were to have positive feelings about animals in general and the more concerned they were about animal welfare. Despite having a different spectrum of pet ownership – none of the Japanese subjects mentioned horses, but they were more likely to have experience with cage birds – the two countries also showed strong similarities: when asked to pick out their favourite family pet, both UK and Japanese respondents named dogs and cats first and second, respectively. Most remarkably, the more pets they had experienced in childhood, the more likely respondents were to have studied a biologically based subject at university.[17]

The evidence points to a robust link between the experience of living with pets during childhood and a person's attitudes later in life to animals in general, including but not limited to pets. However, disentangling cause from effect is not straightforward. Perhaps some other, so far unidentified factor explains the connection. Here's one plausible scenario: some parents, influenced by their DNA and guided by their individual experiences, bring up their children in a way that has a lifelong positive influence on their thinking about animals, one consequence being that both generations are predisposed to keep pets. In itself, pet keeping may be a cultural trait, albeit one primed by genetics, that each generation eventually picks up from the previous one.

AS WITH MOST PERSONALITY TRAITS, when it comes to affection for pets, untangling genetic and environmental factors is difficult. How, for example, do we explain the substantial differences between ethnic groups when it comes to pet ownership? In both the United States and the United Kingdom, whites are more likely than non-whites to keep pets, even after controlling for economic factors. Attachment to individual animals is considerably rarer in rural Africa than in, say, Melanesia.

The differences can be striking where two cultures exist side by side. In the Kalahari Desert in southern Africa, village dogs owned by Ju/'hoansi Bushmen have a hard time of it, most of their physical interaction with people coming in the form of kicks or flying stones (see nearby illustration). They get treated with respect only when they perform their function as part of a foraging or hunting party. In the village, adults and children alike show them no affection and certainly do not treat them like pets.

The Ju/'hoansi find the local Afrikaner farmers' attitudes towards dogs incomprehensible, even contemptible. To supplement their traditional hunting and gathering, the Ju/'hoansi men work part-time on local farms owned and run by the descendants of Dutch settlers. They cannot believe that these farmers feed their dogs on meat while offering their labourers, after a day's work, only maize porridge; they wonder why the farmers rarely invite them across the thresholds of their farm-houses but allow dogs to sleep in their beds. The Ju/'hoansi not only resent being treated worse than dogs but view with disdain the Afrikaners' apparent confusion between canine and human needs.

To the Ju/'hoansi hunters, animals are not people; they are animals, pure and simple, each with a unique identity and behaviours, and should be respected as such. They have their own kind of empathy for the animals around them – but it differs greatly from Western pet owners' understanding of the term. Like many other hunter-gatherers, they can be completely absorbed in the ways of the animals they hunt (the Ju/'hoansi hunt with poison arrows and may have to track their prey for days while waiting for the toxin to take effect), but they are never sentimental. A hunter who imputes human motivations to his prey will likely lose it. Thus, although domesticated, their dogs are still animals, still separate. In their words, 'The problem with you whites is that you think dogs behave like humans or even that your dogs think that they are humans. They are not humans. They are dogs. Their ways are different.'[18]

As modern exponents of Paleolithic lifestyles, the Ju/'hoansi show none of the affection for animals displayed by their counterparts in Southeast Asia and Amazonia. Their dogs are working animals, not

Ju/'hoansi dogs are not treated like pets.

pets, and pet keeping seems to form no part of their culture. Whether such apparent racial differences also reflect genetic factors that influence affection for animals remains a topic for investigation – and there are plenty of other potential explanations based on cultural dissimilarities.

THE APPARENTLY SIMPLE decision to acquire a pet is thus more nuanced than it might seem at first glance. For some, considerations of fashion or status may prompt the initial decision to adopt, with the benefits of companionship coming later as the relationship develops. Such considerations will certainly influence the choice of dog versus cat and the preferred breed of either. Cultural elements undoubtedly come into play, not least in the types of animal available. Family traditions also play a part. But while every pet owner's relationship with his or her animal companion incorporates a slightly different combination of motivations, at its core the bond between them rests on a foundation of affection and, to a variable extent, indulgence.

It was probably ever thus. Despite undeniable differences of detail between the pets that our hunter-gatherer ancestors kept and those we prefer today, there's no reason to suppose that the underlying incentives have changed very much. Cultural influences – imitation of others' behaviour – remain strong, even though the limits of the peer group have changed from the few dozen people living in the same village to the global 'tribes' visiting the same social media sites. The same holds for status. But the affection that ultimately develops must have been a feature of human behaviour for tens of thousands of years and forms a consistent thread running through pet keeping from its unrecorded beginnings right through to the present day.

A desire to keep pet animals seems integral to human nature, transcending culture, race and economics. Of course, the number of people with the time and resources to indulge in this habit has varied enormously over the history of our species. In the relatively egalitarian societies of our hunter-gatherer ancestors, most women and children may have kept some kind of pet animal. The stratification of society that followed the first agricultural revolution around

12,000 years ago made it more difficult for those in the lower ech-
elons to keep pets, apart from the farm cats and herding dogs that
pulled their own weight. More recently, the overall rise in standards
of living have once again made keeping pets possible for most everyone who feels inclined to do so. Affluence has even freed a few to
indulge their pets in hitherto unimaginable ways – but is the largesse
they extend to their companions any more precious or personal than
a young Guajá woman's breast-feeding of a monkey? The underlying
emotional connection may well be the same.

Admittedly, not everyone feels the same way about animals. Genetic factors seem to predispose some people to form a lasting bond
with animal companions, while others remain immune to their
charms. Whereas the minority today who strongly disapprove of pets
on principle (about one in forty people in most Western countries)
tend to focus on the downsides of pet ownership, pet owners tend to
minimize the risks presented by their animals, both to themselves
and to society. Once the relationship with the animal has become
established, most owners display a remarkable degree of loyalty, over-
emphasizing their pet's better points and overlooking their faults.
Just as parents always rate their own children above average (despite
any evidence to the contrary), so do pet owners rate their pets. Plausibly, this rose-tinted view may help to account for the extraordinary
amounts of money that some owners spend on their pets, not to mention the time they devote to them and their tolerance of the inevitable inconveniences that arise – especially with dog ownership. If
pet keeping is part of human nature, such apparent blindness requires
some explanation – as do the almost mystical powers that some,
the popular media especially, seek to confer on animal companions.

Given the ever-increasing rate of change in human societies, the
image of pets (dogs especially) unsurprisingly shows signs of shifting yet again: the mere pleasure that pets undoubtedly bring to their
owners' lives no longer justifies their existence. Towards the end of
the twentieth century, dogs morphed from useful companions into
various kinds of minor superhero. We found many new uses for their
particular brand of intelligence, building on the invention, following

World War I, of guide dogs for blind people: dogs now assist wheelchair users and alert epileptic and diabetic owners to imminent seizures (see nearby illustration). Many people now make even bolder claims about their abilities. Some argue that simply owning a dog can stave off heart disease, draw out withdrawn children, and counteract feelings of loneliness. (In the main, cats have eschewed much of this speculation, as if it were somehow below their dignity.) Although regular pet owners might not have experienced such heroics for themselves, all of this positive press has nevertheless infused pet keeping with a warm, fuzzy glow and inured it a bit more to criticism. But can all these claims have merit, or have those poised to profit most from our ever-increasing expenditure on our pets exaggerated them, as some cynics claim?

Dogs can now be trained to assist users of wheelchairs.

CHAPTER 3

Pets – the Healthy Option?

Half a century ago, most people would have dismissed the idea that simply living with a pet might be good for one's health. Sure, a dog will get you out in the fresh air, and it's pleasant enough to stroke a cat, but anything more than that? Ridiculous. Yet nowadays the press and even some experts widely tout pets as a panacea for all kinds of ills: high blood pressure, stress, loneliness, heart disease and depression, to name but a few. Books with titles such as *Every Dog Has a Gift: True Stories of Dogs Who Bring Hope and Healing into Our Lives* and *Healing Companions: Ordinary Dogs and Their Extraordinary Power to Transform Lives* crowd the shelves of the pet section in bookshops the world over. The Internet yields seemingly infinite articles extolling the virtues of pet keeping (alongside pieces such as '10 Reasons You Aren't Losing the Last 10 Pounds' and '14 Reasons to Have More Sex'). Here are twelve ways pets improve your health (the italics are mine): 'Pets *may* lower your cholesterol; Pets *help* relieve stress; Pets *may* reduce your blood pressure; Pets boost your fitness; Pets reduce your cardiovascular disease risk [so there's a couple they're confident about, then]; Pets *may* prevent allergies in children; Pets relieve depression; Pets ease chronic pain; Pets improve relationships; Pets monitor health changes [save time/money on medical check ups!]; Pets boost your self-esteem; Pets bring your family closer together.'[1]

These extraordinary statements seem to have materialized out of thin air. When I took on my first dog, forty years ago, I'd have been a laughing stock if I'd claimed doing so would make me live longer. Today, while wide-ranging health claims may not be the overriding factor convincing people to take on a pet, they certainly imbue pet keeping with the kind of warm glow that helps those who enter into pet ownership to justify their decision – to themselves, at least, and possibly to others.

Science has struggled to keep up with all these claims. Reliable studies have generally failed to find convincing proof that living with animals makes their owners healthier. Even where evidence does seem to support the idea that people with pets are in better shape than those without, there may be explanations other than pet 'magic'. Our vaunted notions of the extraordinary healing powers of pets suggest that something in our human nature inclines us not only to keep pets but also to see them through rose-tinted spectacles.[2]

Beliefs in the healing power of pets have spread remarkably quickly and widely. At one level, we can view 'pet therapy' as part of the recent movement towards complementary and alternative medicine, but looked at dispassionately, pets would seem improbable therapeutic agents. They harbour all manner of microorganisms, including some that can make people sick. Their behaviour is often unpredictable: dogs seem to delight in licking the faces of those who least appreciate this gesture of affection, while cats have their own rules about whom they will allow near them. Pets seem altogether too wild and capricious, especially by comparison with the beguiling packaging of a homeopathic medicine or the soothing hands of a reflexologist. That we would ever consider pets for the role of 'co-therapist' may say more about the instinctive appeal they hold for us than about any tangible benefits they may confer.

REMARKABLY, WE CAN trace this tide of positivity towards pets to just two early studies, one, a series of case histories conducted by a psychotherapist, and the other, a study by a team at the Universities of Pennsylvania and Maryland.

Dr Boris Levinson, a psychotherapist based at New York's Yeshiva University, coined the term 'pet therapy' around 1960, after he stumbled on a role for animals as 'mediators' for uncommunicative patients, specifically children. As the story goes, one day a mother and her chronically withdrawn son arrived early for their appointment to find that Dr Levinson had not yet removed his dog Jingles from his office. The child, who had never spoken during previous sessions, suddenly began to address the dog and then began to respond to Dr Levinson's questions, albeit directing his replies not to him but to the dog. Inspired by this success, Levinson began to incorporate Jingles into his treatment of other young patients and found that many happily played with the dog even when they were uncomfortable speaking to him: by gradually insinuating himself into these games, he found that he could establish trust with many of these children. While his academic colleagues were initially sceptical about using animals in psychotherapy, he energetically promoted his methods, which, by his death in 1984, had become widely accepted in the United States.[3]

Although perhaps the first to advocate using dogs as therapeutic aids, Levinson was not the first to notice that they can comfort people with mental disorders. In the United Kingdom, the York Retreat, founded by the Quakers in the late eighteenth century, was the first asylum to include contact with domestic animals as one element in the provision of a homelike environment for its inmates (mostly schizophrenics). Sigmund Freud himself often allowed his Chow Chow Jofi to attend his therapy sessions (see nearby illustration), claiming that the dog could sometimes better assess his patients' state of mind, but he apparently did not consider elevating Jofi to the status of co-therapist, as Levinson later did with such success. Even at its strongest, though, Levinson's advocacy of the benefits of contact with dogs did not extend beyond children, although he did suggest that all children, not only those with psychological disorders, could improve their emotional development by caring for animals.

A second study, some twenty years after Levinson's, opened the way for a much more general claim about the benefits of pet keeping: that pets could bring about an across-the-board improvement in cardiovascular health. Pioneering anthrozoologist Erika Friedmann and

Freud and Jofi.

colleagues examined how the home lives and psychological states of heart-attack victims might affect their survival rates after discharge from hospital. They found that, as expected, those with more severe cardiovascular symptoms were the least likely to be alive one year later, as were those who lived alone. The surprise came when she examined the numbers of survivors who kept house pets: 40 per cent of those without a pet had died, but over 94 per cent of those with pets had survived. Thinking that this might reflect dog owners' need to take exercise, she removed them from the analysis and found that of the ten people with 'other pets', precisely none had died.

Friedmann's study triggered a search for explanations, one that continues to this day. How could the simple companionship of an animal like a cat or a rabbit facilitate the recovery of a severely dam-aged human cardiovascular system? One answer might simply be that it doesn't. The original study was small – including less than a hun-dred patients – and after its publication other scientists claimed that the average younger age of the pet owners in the group might explain some of the apparent associations between pets and health. When Professor Friedmann repeated her study on a much larger sample of 369 survivors of heart attacks, the dog owners did well (only 1 in 87 died), but cat owners were, if anything, more likely to die than those with no pets at all. Subsequent studies have tended to support the association between dog ownership and healthy hearts but have failed to disentangle cause and effect – people who are unhealthy to begin with may rarely choose dogs as pets because they know they'll have to walk them. Furthermore, ownership of other pets makes little differ-ence either way (indeed, more than one study has shown that those who own cats are more at risk of a second heart attack than those who don't). Nevertheless, pet owners, the media, and even medical practi-tioners seem only too keen to perpetuate the myth that pets improve health. (In one recent survey conducted in the United States, 97 per cent of doctors polled stated their belief that pets have health benefits, and 74 per cent stated that they would 'prescribe' a pet to improve overall health). Such misplaced enthusiasm cries out for explanation.[4]

BORIS LEVINSON WOULD probably be more than a little surprised by the extensive application of his claims for the therapeutic powers of animals. The UK-based charity Pets as Therapy has registered over 28,000 dogs as therapists since its founding in 1983 and currently has over 4,000 dogs (and a few cats) on its books. The Alliance of Therapy Dogs, which operates in the United States and Canada, has registered over 14,000 animals. Dogs, though widely used, are by no means the only species deployed: the US-based Pet Partners organization registers a variety of animals for use in therapies, in-cluding dogs, cats, rabbits, birds, guinea pigs, rats, miniature pigs, llamas, alpacas, horses, donkeys and mini-horses. The enthusiasm is real enough, but are these therapies always effective? And for those that are, how does the animal make its contribution? Or does success depend on the motivations of the therapist?[5]

We should distinguish here between animals used simply to brighten someone's day and those employed as part of a treatment program specific to one individual. Some who practise the latter object to references to the former as 'therapy', which they prefer to restrict to interventions tailored to help a specific patient. They claim that we should refer to the now common practice of taking dogs, rabbits and other animals into long-stay hospital wards and homes for the elderly as 'animal-assisted activity'. In general parlance, however, the word 'therapy' widely applies to both: almost all dogs registered by Pets as Therapy simply visit the elderly and infirm, with no individually tailored treatment involved. Also, both practices, designed to improve a person's mental state, differ from the help given by assistance dogs (guide dogs and hearing dogs, for example), which primarily provide physical aid – which is not so say that they don't offer psychological benefits as well.

We can hardly regard dolphins as pets, but the extraordinary enthusiasm about their healing powers perfectly illustrates just how powerful such beliefs can be. Over the past twenty years, so-called dolphin-assisted therapies have proliferated worldwide. Some simply consist of supervised swimming with wild dolphins; others entail more structured activities involving captive dolphins. Their proponents claim that these encounters can alleviate or even cure a wide range of conditions both mental and physical, including AIDS, cancer and clinical depression. Some even assert that the dolphins actively take part in the therapies. Here's a sample: 'Very often the therapists tell me of their experience of working with patients in their office. They devise their plans for the day, only to find that the dolphins already know what needs to be done when the therapist and patient come to the water.'[6]

It is widely asserted that dolphins are especially valuable as co-therapists for special-needs children with autism, Down's syndrome and cerebral palsy, and even those in waking comas: parents, desperate to see any improvement in their children's condition, pay as much as

$7,000 for ten hours of dolphin therapy, far more than they would for other animal-assisted therapies (many of which are free or offered at cost). Dolphins supposedly do something 'magical' to the water around them, lifting anyone who swims past into a euphoric state. Maybe it's the sonar that they emit (the clicks dubbed into TV documentaries about cetaceans). Maybe it's their alleged ability to trigger in us a meditative state or to 'synchronize' the left and right hemispheres of our brains. (Or maybe it's just fun to swim in warm water, in an exotic location, alongside a playful mammal.)

Whatever the claims, numerous attempts to nail down the effects of dolphin-assisted therapy have failed. Many of the published studies have come from the therapists themselves, and none so far has stood up to the standards demanded of medical trials. For example, although some researchers have concluded that the ultrasound emitted by dolphins could theoretically affect human tissues, the amount typically received falls far short of that known to have any effect.[7]

Worse, regardless of the efficacy, it seems highly unethical to keep highly intelligent marine mammals in captivity for the purpose of human therapy. The dolphins themselves sometimes display their disquiet by attacking those who work with them and, on occasion, the patients they are supposed to heal. As a consequence, the Whale and Dolphin Conservation Society has called for a worldwide ban on dolphin-assisted therapy, citing the dearth of proven benefits and lack of concern for the welfare and conservation of the animals involved. The organization Research Autism does not recommend dolphin therapy as a treatment for people with that disorder.[8] Indeed, the use of any wild or captive animal as a therapeutic aid raises serious ethical questions since we cannot know how the animal perceives its situation. Domesticated species, in contrast, have by definition become adapted to man-made environments and so can better interpret human behaviour; moreover, their trainers can respond appropriately and sensitively to their needs.

Some also make extravagant claims for the therapeutic value of horses. Several types of therapy use domestic horses, and equine enthusiasts often claim both physical and psychological benefits for any one-on-one interaction with a horse. Therapeutic horseback riding can improve postural control, balance and reflexes, but some assert it also improves self-confidence and emotional control. As its name implies, equine-assisted psychotherapy focuses on the psychological benefits, using the horse more in the role of co-therapist, as envisaged by Levinson. Self-help TV shows have widely promoted such therapy, and some therapists have made claims about the horse's capabilities, such as that research has proven horse therapy effective for people struggling with issues such as depression, addiction, eating disorders, and recovery from trauma or abuse. A recent assessment of fourteen studies of equine-related treatments for mental disorders concluded, however, that current evidence does not support these assertions.[9]

The psychological benefits of interacting with dolphins and horses thus seem largely the product of wishful thinking on the part of both therapist and client. Can the same be said of pet-assisted therapy? Most therapy pets are dogs, so not surprisingly they are the best studied. Over the past two decades, dozens of investigations of dog-assisted therapies have been published, and the large majority have found that the therapy was effective for the majority of patients.[10]

So why the scepticism? Because it's very hard to pin down how a complex procedure such as pet therapy might work. First there are the expectations of the patient or, in the case of a child, the parents. The supposed health benefits of pets have received so much publicity that simply expecting the therapy to work might bring about an impression of improvement. Second, we know that placebo medications can, in some cases, achieve many of the same results as real medications (simply because patients believe the pills are working), and that the placebo effect is particularly prevalent in treatments aimed at mental disorders. Could this not be true also of pet-based therapies? No doubt the therapy dog will be novel, exciting and energetic, all of which may lift the patient's mood enough to look like improvement – but this effect lasts only as long as the dog is present. Third, the enthusiasm of a human therapist convinced of the efficacy of the dog may rub off on the patient – thereby temporarily raising his or her mood.

What evidence suggests that the pet itself enhances the mood of the patient? In the early days of dog-assisted therapy, my students and I set up therapy sessions for children with Down's syndrome. In some sessions, the therapist encouraged them to interact with a black Labrador retriever; in others she offered the same games, but with a plush-covered stuffed-toy simulacrum we named Max. Max wasn't only good to stroke and hug; he had a collar and lead and so could be taken for 'walks' (drags?), as well as a gullet so he could be 'fed' biscuits. We offered no specific therapy, since we were only interested in whether the real dog elicited a change in the children's behaviour – and it did. Compared to Max, the real dog held the children's attention for much longer, and they were also more attentive to the handler when the dog was real.[11]

The advent of more realistic robotic animals has somewhat undermined the hypothesis that only a live animal can provide therapeutic benefits. Studies have shown both AIBO (an obviously mechanical robot dog) and PARO (a fluffy and remarkably realistic – and cute – harp seal pup; see nearby illustration) to be as effective as dogs in alleviating loneliness in residential care facilities for elderly people. Thus perhaps interaction with something responsive – but not necessarily alive – produces most of the benefits. (Research into robot-assisted therapy is proceeding apace in Japan, which faces an acute shortage of people to care for its large and growing population of elderly people, many of whom consider dog hair and saliva repellently unhygienic).[12]

Assessing the efficacy of dog-assisted therapy entails another problem known to scientists as the file-drawer effect. It's a little-appreciated fact that much scientific research stays locked in the computers of the people who did it. This isn't always a bad thing: for instance, someone who initiated a study may have realized halfway through that he hadn't designed it properly; sometimes a well-designed experiment simply doesn't yield particularly exciting results.

Robotic 'pets' can alleviate loneliness.

Science journals aren't as sensation hungry as tabloid newspapers, but they are still generally reluctant to devote pages to 'doesn't show much' studies, so hard-pressed professors understandably put their efforts into writing up their most conclusive results.

Thus studies that show big effects for an intervention – be it a therapy dog or an antidepressant pill – get published, while those that showed smaller or no effects tend to get mothballed. As a result, the scientific literature inadvertently exaggerates the apparent effect of any given intervention, as opposed to doing nothing at all. The reality emerges on the rare occasion when scientists go to the trouble of digging up unpublished studies – doctoral theses, for example – and compare them to the published work. When researchers did this for animal-assisted therapies, they found that the effect in the unpublished studies was about half that in the published ones. Anthrozoologist Hal Herzog commented, 'If 100 people were to get animal assisted therapy, roughly 9 of them will be better off for the experience while the other 91 would have done just as well by staying home and hanging out on Facebook.'[13] (I don't think he meant that Facebook is a viable alternative to therapy for the psychologically vulnerable.) It's worth noting, however, that the same bias would be at play if dog-assisted therapy were found to be harmful under some circumstances – science journal editors love a good controversy. But no such study has surfaced, meaning that dog-assisted therapy is generally benign, even if often only marginally effective.

PET-BASED THERAPY does seem to have real, positive effects with autistic children. Dogs seem especially useful in drawing children with autism-type disorders into conversation with their therapists. While they struggle with eye contact and conversation with humans, many such children appear far less inhibited when it comes to animals – indeed, many show much more interest in pictures of animals than of people. Certainly their parents often regard animal therapy as worthwhile: almost one in four have tried it, and half of them persist, believing it has made their child more sociable (only diet supplementation is more popular). Taken as a whole, studies show that many autistic children do react positively to dogs. Not only do they experience less stress and smile and laugh more frequently, but they also become more conversational with the humans in the room. As in Levinson's original case, the dog can act as a bridge between child and person.[14]

However, not all autistic children respond well to dogs: about a third are indifferent, and a few find the dog's presence upsetting. Much may depend on the dog's training and how well it copes with the child's sometimes unpredictable behaviour. Also, a few atypical children may find a dog and an unfamiliar person about equally threatening.

Dogs may be too bouncy for some autistic children, and less responsive animals such as guinea pigs have recently produced more reliable improvements, especially in classrooms. Many children on the autism spectrum, although educated in mainstream schools, find it difficult to integrate with the other pupils. Compared to otherwise identical sessions using toys (cars, dolls, clay and spinning tops), experimental sessions focused on a guinea pig were much more effective in getting autistic children to socialize with their peers: with the guinea pig they laughed and smiled more and frowned and cried less. Compared with other studies involving adult patients, the children were less likely to behave according to preconceived notions of the 'magic' powers of animals and more likely to respond directly to the animals. Strong indicators suggest that the therapy is effective on a biological level: when the researchers measured the children's skin conductance, which indicates excitement and/or frustration, the autistic children's conductance was generally higher than the other children's but went down when they were handling the guinea pigs.[15]

Formal therapy sessions may not be the most effective use of pets in bringing autistic children out of their shells, since they are usually brief and don't allow the child to get to know the animal particularly well. Obtaining a pet specifically for the child, if done correctly, may allow a more spontaneous relationship to develop and also enables experimentation with other types of animals. Though few and far between, 'therapy cats' can, anecdotally at least, be very successful pets for autistic children (see nearby illustration). Iris Grace Halmshaw, an autistic child considered an artistic prodigy in England, took solace only in painting Monet-style landscapes until a Maine coon kitten called Thula drew her out (according to her mother's account).[16]

A well-documented case study also supports this idea. At eighteen months, Fraser Booth cried incessantly, comforted only by a strip of plastic with which he had formed an inexplicable bond. Diagnosed with autism spectrum disorder, he also lacked muscle tone. By the age of three, he was socially isolated, ignoring the other children at his playgroup and panicking whenever he saw a stranger; moreover, he treated his mother, in her words, 'like a piece of furniture'. Then his parents adopted an abandoned kitten that they named Billy. Within a few days, they saw a dramatic change in Fraser's behaviour. He was far less tense, and as the months and years progressed, he became able to engage with his parents and even with visitors to the house. Billy may not be particularly magical, although he seems more flexible than other cats, but Fraser thinks he's very special. As he said in an interview – a remarkable event in itself, considering his former aversion to strangers – 'I like hugging him and kissing him and jiggling him around.'[17]

Billy's arrival in the house when Fraser was three may have been the key to the success of their relationship. When Fraser was born, his parents already had a middle-aged cat called Toby, but Fraser and Toby showed no interest in one another. One study of autistic children in France suggests that this phenomenon may be general: while existing pets seem to benefit autistic children little, pets obtained during mid-childhood can help them become more sociable – perhaps because the arrival of the animal breaks into the child's self-absorption.[18]

Adding a cat to the household can have positive effects on autism.

ADVOCATES ALSO ENTHUSIASTICALLY promote pet-assisted therapy as beneficial for the elderly; many practitioners claim that it helps bring geriatric patients back from a life of increasing isolation. More and more older people live in purpose-built accommodation. As recently as thirty years ago, few such establishments permitted animals, considering them unhygienic and too much trouble to care for. Nowadays, however, many geriatric hospitals, old people's homes, and day centres welcome visits from the legion of enthusiastic pets-as-therapy teams. Some facilities even have resident pets. Contact with animals may help to alleviate loneliness and depression among elderly people and also reduce some symptoms of dementia.

The many positive signs they observed encouraged the pioneers of such pet visitations. In 1990 psychologists Val Elliott and Derek Milne recorded visits by Thistle, a Wheaten terrier, and her handler to two wards in St George's Hospital, Morpeth, in northeastern England. The twenty-seven residents of Ward 1 suffered from dementia: some were long-term patients; others were there to give respite to their usual carers. In Ward 2, most of the twenty-one patients were depressed. In both, the patients became much more animated when Thistle was present, interacting more not just with the dog but with the staff and their fellow patients. The nursing staff generally agreed that many of the patients seemed better both during and after Thistle's visits, and none could identify any patient whose behaviour had deteriorated as a result. Moreover, many of the staff reported that the dog's visits had raised their own moods and improved their morale, creating a better atmosphere all round.

As professional psychologists, Elliott and Milne were rightly sceptical about whether the dog was the magic ingredient in the therapy. The changes in the patients' behaviour were more marked in Ward 1 than in Ward 2. Rather than attributing this difference to the different disorders on the wards, they suggested that it might have stemmed from the enthusiasm for dogs shown by one of the nurses in Ward 1. It was not too far a stretch to suggest that the main effect of the dog might have been not so much to raise the mood of the patients directly as to change the way the staff behaved towards them, and Elliott and Milne raised the possibility that 'the same results could have been achieved by these people speaking to each patient without the dog.' As they pointed out, however, 'Whether or not the dog was the prime contributor to the beneficial results, without the dog there would be no therapy "package".' They also highlighted the contribution such visits might make to the 'normalization' of the otherwise sterile and alien ward environment, bringing back a touch of the patients' everyday lives before their hospitalization.[19]

This normalization evidently extends as much to staff as it does to patients, sometimes catching them by surprise. Debra Fila, working as an administrative head nurse in an acute care unit at the University of California Medical Center in the mid-1980s, was shocked by the reaction of her colleagues to the admission of a patient for reconstructive eye surgery accompanied by his guide dog, a yellow Labrador named Woo-Woo. 'For me, what could have been a nightmare, turned into one of the most satisfying experiences of my nursing career. I had no idea that the nursing staff, patients, families and physicians would react in a positive way to a large dog living on the hospital unit . . . While Woo-Woo sat behind the nursing station, I observed physicians and nursing staff stopping to pet her, smiling and saying a quick, friendly "hello." Comments from the staff indicated positive feelings and an atmosphere of reduced stress.'[20]

Not all elderly people want contact with pets, so pet therapy is unlikely to be universally beneficial. Unsurprisingly, patients' preferences usually relate to their previous experiences with animals: in one study of residential care establishments in southern Mississippi, none of the 12 per cent who declined dog-assisted therapy had kept pets prior to going into care, whereas almost all of those who accepted the therapy had. In this population, visits by a dog often prompted spontaneous recollections of pets from decades before and the corresponding attitudes towards them. One resident 'remembered fondly how her pet dog would bring dead squirrels, rabbits and opossums back to her', which she was in the habit of filleting and then frying as her next meal. (Presumably she allowed the dog the remainder of the carcass.)[21]

The medical profession as a whole, however, has been rather resistant to pet therapy, perhaps seeing it as yet another manifestation of alternative medicine, and in-depth investigations of its efficacy have been few and far between. In one such study conducted in a nursing home in Rome, the scientists examined the reactions of dog-loving residents to regular half-hour sessions with one of two therapy dogs, a five-year-old golden retriever and a six-year-old cocker spaniel. Playing with the dogs brought about a remarkable change in their behaviour: from being largely apathetic they gradually became not only more active and animated but also more spontaneous with the dogs – for example, throwing a ball for the dog to retrieve. Another notable feature of interaction with dogs is the extent to which the elderly residents reached out and touched them, reducing their feelings of emotional isolation.[22]

Such increases in interaction should – as most of the studies demonstrated – lighten the residents' moods and reduce their feelings of depression, at least temporarily. We don't know whether these effects are ephemeral or longer lasting, but anecdotally residents do not typically lose interest in the dog's visits over time as can happen with other, less engaging kinds of therapeutic activity.[23]

Perhaps the greatest challenge facing those who care for elderly people today is the 'global epidemic' of dementia. To a degree unprecedented in human history, bodies are outliving minds. Estimates put the number of people worldwide living with dementia at 50 million, with a total cost to the global economy of over $600 billion each year. Effective treatment with pharmaceuticals has remained elusive – in the past fifteen years, only three new drug therapies have obtained approval to treat Alzheimer's disease – and much research focuses on reducing the impact of dementia on its sufferers and their carers.[24]

Given the enthusiasm of those who promote pet therapy for elderly people, claims that contact with animals ameliorates the symptoms of dementia are unsurprising. The small number of formal studies done do not indicate that pet therapy has a major effect on cognitive function (that is, it does not halt the progression of the impairment). But studies do show that pet therapy can reduce the agitation of patients suffering from dementia and, as with those suffering from depression, enhance their interactions with their carers. Overall, regular contact with a therapy animal can enhance the dementia sufferer's quality of life, as measured both formally and in observed exchanges between residents and dog handlers. As with depressed elderly people, these conversations often involve residents' recollections of the pets they owned when younger and the emotions those pets evoked.[25]

Continued contact with pets can bring a nostalgic joy to people who can no longer care for a pet of their own. As such, pet visitation programs seem worthwhile even if, as seems likely, the medical benefits are marginal and may be mostly imaginary. Even when pet therapies are effective, the pet's precise role is rarely clear. Thus we may feel tempted to dismiss animal-assisted therapies and lump them together with other 'alternative' and 'complementary' remedies of dubious efficacy. However, the involvement of the animal does genuinely seem to galvanize a change in the behaviour of human participants. The increasing use 'pet therapies' points to psychological mechanisms potentially unique to interactions between individual people and animals. Whether they most directly affect the enthusiasm of the therapist or the mood of the subject remains unclear – and possibly difficult to determine in a therapeutic situation. Perhaps we can uncover less ambiguous evidence by examining claims that pets affect the way people feel in their everyday lives.

THE VERY PHRASE 'animal companion' suggests that pets should effectively combat loneliness, and this is widely believed to be the case. However, is pet ownership a protection against or a response to loneliness? Do people who crave but cannot get enough human company turn to animals as the next best thing, and is this salve effective? The answer seems to be sometimes and – for some as yet unknown reason – mainly among women.

Loneliness is a special problem for older people who remain sufficiently well to live in the community but whose social circles are diminishing. In western Europe and the United States almost half of those over age sixty-five consider themselves moderately or severely lonely. Loneliness is a subjective concept and varies not just from one individual to another but also between cultures. For example, in Greece, where most elderly people live with or very close to their families, many still report feeling lonely: loneliness seems to arise when one's social life does not match one's expectations, rather than emerging automatically when the number of social contacts falls below some arbitrary threshold. We must therefore examine how individuals' expressions of loneliness change with their circumstances rather than, for example, simply comparing loneliness among those who do and don't have pets.

The English Longitudinal Study of Ageing, which since 1998 has tracked more than 5,000 individuals over fifty, provides unique and surprising insights into the complex relationship between pet keeping and loneliness. About 40 per cent of subjects owned pets at the outset, declining to 30 per cent over the next ten years, presumably due to a combination of reduced income, deteriorating health, and concern that a new pet would outlive them. By tracking who owned pets at each stage of the study and comparing this with how lonely each person felt, the investigators could identify which pets subjects had acquired to stave off loneliness and which had protected their owners against it.[26]

Among the men (almost 1,000 of them) researchers could discern no relationship between pet ownership and loneliness. Among the women, the pet owners had started out lonelier than the rest, but many became less lonely over the following ten years. Thus for women in their fifties, pet ownership may be an attempt to combat loneliness, possibly due to loosening ties with their children; furthermore, it seems to work, at least for some. Neither feelings of loneliness nor the expectation that a pet might cure them seem to play a major part in why men in the United Kingdom obtain pets in late middle age.

If pets stave off loneliness, they may be saving lives, because lone-liness is a killer. Lonely people are more likely to have heart attacks, to commit suicide, and to drink excessive quantities of alcohol. Con-sequently, loneliness may be up there with cigarette smoking and obesity as a health risk.[27] Loneliness is often thought to have two causes: lack of a network of people to call on for help – relatives, friends, colleagues – and lack of a close one-to-one relationship such as with a romantic partner. It's not entirely clear where a pet might fit into this framework, and different species, even different individual animals, might play different roles. A well-behaved dog might con-ceivably help with both, bringing its owner into contact with other people during walks and also acting as a 'confidante' if the owner lives alone. Cats and small pets provide companionship at home but are unlikely to extend the social networks of their owners, apart from the tiny minority who take their animals to pet shows.

Older people, sometimes lacking the opportunity to develop new relationships with other humans, may turn to pets as a substitute, especially if they can recall satisfying relationships with household pets from their earlier years. Elderly women who keep pets become more attached to them the lonelier they feel, suggesting that they indeed turn to their animals to compensate for a dearth of human relationships. However, these attachments seem to have little effect on their general health, throwing into question whether their pets are genuinely having an impact on their loneliness – or at least those aspects of loneliness that most affect health. Perhaps the explanation lies in most elderly women's choice of pet, cats, which don't provide an incentive to leave the house and interact with other humans in the way that a dog would.[28]

Although pet therapy seems effective in alleviating the signs of depression in institutional settings, clear evidence that those with household pets are any less depressed than those without has proved elusive. Indeed, at least one study, conducted in Norway, found that among elderly women, cat owners were more prone to depression than those who eschewed the company of felines, whereas dog own-ers were no more or less depressed than the majority of women who had no pets. The same did not hold for the men in the study, who were universally more depressed than the women – perhaps less able to cope with the Scandinavian winter. In the United States, unmarried men of all ages who owned pets reported being more depressed than those who had no pets, apparently regarding their animals as a burden rather than a benefit. Another study, which indicated that only unmarried women felt the benefits of dog ownership, confirmed this connection. However, a fourth study, conducted in Australia and looking mainly at married women who owned dogs, highlighted high levels of depression among those who spent the most time with their pets. Their strong attachments to their animals seemed potentially unhealthy: some felt unsure whether they could leave their pets if they had to undergo surgery.[29]

Female dog owners may, however, find genuine consolation in the company of their pets following the death of a spouse. Typically, recent widows find their health deteriorating unexpectedly, but possibly less so if they own a dog. The dog seems to be a special help only if the person has few others to confide in, so even here the benefits of dog ownership may be somewhat restricted: perhaps dogs can't substitute for people, but if there's no one around, they can be the next best thing.[30]

If pets seem to offer little benefit to elderly people in staving off depression, except perhaps in institutional settings, then dogs do at least have the potential to enhance their social lives. In one early study, residents of trailer parks in Sacramento, California, recorded their conversations during the course of everyday walks. Some were dog owners, while others were not; the dog owners went on walks accompanied by their dogs and also alone. Dog ownership did not affect the number of conversations subjects had while out walking, but it did influence what they discussed. Passers-by talked to the dog owners about their dogs (whether the dog was there or not), and these conversations, even once they'd strayed onto other topics, focused on present-day happenings and feelings: 'Tiger hasn't been eating so well the last couple of days. He's been turning his nose up at his dinner. Maybe it's the change in the weather. I know he gets moody, and I get a little bit moody too.' The dogs seemed to steer attention to the here and now, providing participants with an alternative to their usual topics of conversation, which, perhaps unsurprisingly, given their age, often focused on health issues.[31]

A COMMON THREAD running through all these studies is the way dogs bring people together. Since social contact is beneficial to many aspects of health, this may account for observed therapeutic benefits of dogs as well as for the almost complete ineffectiveness of cats, which prefer to interact one-to-one and in private. The small number of studies examining the role of dogs as social bridges confirm that there is, indeed, something rather special about this species, even though the precise reasons why most of us react positively to dogs remains unclear.

As actively sociable animals themselves, dogs try to engage with people they don't know (in addition to people they do) and hence can catalyse a conversation between humans – even if the exchange is limited to 'I don't like dogs. Please get him away from me!' 'Oh, I'm so sorry. BUSTER, COME HERE.' But is dogs' natural nosiness the only factor at work, or does their mere presence change people's perceptions? Would any non-threatening animal have the same effect?

In one brave study conducted in California, a young woman sat in a park, reading and making occasional jottings in a notebook, accompanied by either a rabbit or a turtle; or instead of bringing an animal with her, she blew bubbles from a wand every now and then. The bubbles tended to attract more passing children, but an equal number of children and adults approached her when she brought the rabbit. Even the turtle elicited some enquiries. Something about animals – even unobtrusive ones like rabbits and turtles – seems to break down barriers between people.[32]

Unsurprisingly, dogs exert a powerful attraction. Those that act as guides for blind people, although trained not to initiate interactions with the people they encounter, nevertheless elicit a great deal of attention from passers-by. Moreover, people who have not previously met the dog's handler seem most powerfully attracted. (This was a major difficulty for those relying on guide dogs in Japan: young people, especially, seemed drawn like moths to a flame to touch these dogs, and few appreciated that both the dogs and their blind partners can find such attention highly distracting.)

Even scruffy or potentially aggressive dogs seem capable of overcoming people's natural reticence when encountering strangers. In one study, a handler dressed in torn, dirty jeans, scuffed work boots, an old T-shirt, and a stained donkey jacket. His dog wore a studded collar with a frayed piece of rope for a lead. Nevertheless, passers-by were eight times more likely to approach or smile at the handler than when he was on his own. Only about 35 per cent more people interacted with him if both he and the dog were dressed smartly. Another study compared reactions to a Labrador and a Rottweiler to test whether the breed of dog had any impact on how approachable a person became. The length and type of interaction varied – for example, the Rottweiler did not draw people into conversation to the extent that the Labrador did – but any dog was better than no dog. Moreover, people remember that a dog accompanied someone on a previous occasion and are more likely to engage in conversation on subsequent occasions, even if only to ask, 'Where's your dog today?'[33]

The ease with which dog owners strike up conversations with one another seems to reflect an aura of trustworthiness endowed by the dog that that owner would not possess if alone. This effect seems remarkably powerful: for example, in one study conducted in France a twenty-year-old male delivered the following chat-up line to 240 young women chosen at random while walking alone in a pedestrian-only area: 'Hello, my name's Antoine. I just want to say that I think you're really pretty. I have to go to work this afternoon, but I was wondering if you would give me your phone number. I'll phone you later and we can have a drink together someplace.' When Antoine was accompanied by a dog (not his own), almost a third of the young women handed over their phone numbers; when he was on his own, fewer than 10 per cent did so (see nearby illustration). In another study, adding the phrase 'with a dog' to the dating profile of a man

Spot the difference – a dog can make a man seem more trustworthy.

clearly interested only in short-term relationships nevertheless induced some women to rate him as a serious marriage prospect. The
power of dogs to draw people into conversation suggests that a facility with animals is intrinsically appealing (even when that facility
has been faked). Even cats, despite their solitary natures, may possess
some of the same properties: one study found that they made men
(but possibly not women) seem more trustworthy.[34]

Pets, especially dogs, may therefore play a significant role in communities and provide benefits for people of all ages, not just the elderly. They can catalyse casual conversations that may develop into
more lasting relationships. More generally, owning a dog (but not a
cat) requires regular excursions to local open spaces, producing opportunities for acquaintance with neighbours who might otherwise
have remained strangers. The demands of pet keeping can establish
informal aid networks, even one as simple as two cat owners feeding
each other's animals while one is on holiday.[35]

THE SOCIALIZING EFFECTS of pets – or at least dogs – seem robust,
despite the wild exaggeration of some of their more specific properties.
What of the apparent effects on heart disease? Spurred on by Professor Friedmann's original study in the late 1970s, scientists all over the

world began to explore whether pets might serve as a reliable buffer against the condition. Well-established mechanisms already existed to explain why giving up alcohol (for example) might prolong life, but not why obtaining a dog or cat might do the same. One avenue of investigation showed initial promise: high blood pressure is a good predictor of a heart attack, so if interacting with animals reduced blood pressure, that might explain the 'pet effect'. And indeed, contact with (calm) animals does seem to have a calming effect, reducing both heart rate and blood pressure during mildly stressful activities. Even watching a tankful of fish or having a dog in the room will do the trick (then again, so will watching a video of the same). Stroking the dog yields an even bigger effect.[36]

How these short-term changes might routinely translate into longer-term improvements in physical health is much less obvious. The quiet moments spent stroking and talking to our animal companions, while probably our most pleasant interactions with them, are by no means the only ones. It is entirely possible that the stressors associated with owning a pet – when it misbehaves, when it gets injured – in combination with the financial demands, more than cancel out the stress-busting moments of the relationship. Furthermore, since petting is the most calming part of owning a pet, the highly tactile cat should be more of a boon to health than the dog – but the data indicates the opposite.

Following the original suggestions that pet owners might live longer following heart attacks, the search widened to examine the more general health of pet owners. Initially, the signs were promising for the supporters of the 'pets for health' hypothesis. A large study in Australia suggested that pet keeping might be as protective as switching to a low-salt diet or cutting down on alcohol. In England, new pet owners rated their own health as improved for the first few months after acquiring the pet (but not, perhaps tellingly, a year later). Since then, studies showing no overall difference in health between those who own pets and those who don't, or even an association between pet keeping and poor health, have balanced out those that show a positive effect. In Australia, scientists from the appropriately named

Black Dog Institute in New South Wales surveyed a large cohort of cardiac patients and found that the cat owners were more likely to be rehospitalized or die within the following year than the patients without pets. Dog ownership produced little benefit either way. It seems increasingly likely that the so-called health-promoting powers of pets are back-to-front: pet owners tend to be more affluent – and therefore healthier – than non-owners.[37]

Research has explored potential links between the psychology of pet ownership and the putative health benefits. If pets' capacity to bring people together accounts for many of the psychological benefits that they confer, could this also serve as the basis for claims regarding their effects on physical health? At one time scientists believed the two were interlinked, that the psychological benefits of social interaction induced the physical benefits. People isolated from their fellow human beings face an increased likelihood of an early death, and not simply because no one is nagging them to take care of themselves – eat well, exercise – although this does seem to be a health benefit associated with marriage. The outcome seems to have more to do with a basic human adaptation to rely on others for protection. When we're on our own, we can't help worrying unnecessarily about imaginary threats, probably part of a mechanism that evolved in our ancestors as a way of protecting ourselves from predators when literally no one was watching our backs. Feelings of isolation cause our blood pressure and our stress hormones (including cortisol) to rise, which in turn reduces the effectiveness of our sleep and hampers our immune systems. If they persist over a long period, these factors appear to hasten the day we shuffle off our mortal coil. While strong human networks provide robust protection against premature death, however, strong relationships with pets have no effect at all on longevity. Unfortunately, although pets do provide their own form of companionship, their presence in the home, according to current thinking, has little effect on their owners' lifespans.[38]

Others claim that all these studies have looked in the wrong place. If dog owners have better cardiovascular health than other people, including owners of other pets, then the obvious explanation requires no invisible psychological mechanism: most dogs get at least the occasional walk. In Western societies, only a minority of people get the amount of exercise recommended for good health, but dog owners get more than most. Despite some suggestion that people who choose to own dogs are already more active than those that don't, obtaining a dog usually leads to a small increase in activity – for example, in one study in Western Australia, recreational walking increased by twenty to thirty minutes per week after the arrival of a dog in the household.

Unfortunately, this extra exercise doesn't translate into lower obesity or a longer lifespan. The link between obesity and exercise, we now know, is something of a myth: being overweight is largely the consequence of overeating. More disappointing is the apparent lack of a link between walking the dog and life expectancy, since according to one recent estimate just over an hour of moderate exercise per week extends life by almost two years – a remarkable 'payback' of around five to one (that is, every hour of exercise should, for the average middle-aged person, delay death by five hours).[39]

Many of the people dog walking in the park next to my home don't seem to walk very fast – and these days many seem more interested in their phones than their dogs – so perhaps only a minority cross the threshold into health-promoting 'moderate' exercise. Nevertheless, exercise has other benefits, including preventing or delaying heart disease, type 2 diabetes, and some cancers, so the brisker dog walkers are almost certainly enhancing their health.

Given that pets' health-promoting powers seem largely illusory, we must dispel one final fallacy: that pet ownership lowers the cost of health care. Several studies show that pet owners make fewer doctors' appointments than people with no pets, but cause and effect are unclear. Common sense would suggest healthy people have the time, motivation, and spare cash to devote to a pet (and the physical energy to take on a dog), while unhealthy people have less of these; hence the latter don't burden themselves with (any more) pets but do make more visits to the doctor. However, while pet ownership might (optimistically) be associated with an estimated $12 billion annual reduction in health-care costs in the United States, the cost of veterinary care alone, never mind the other costs incurred, would more than cancel out any savings. For countries such as the United Kingdom, where tax-payers pick up the tab for health care, it seems somewhat invidious to suggest that people should take on the considerable costs of own-ing a dog (estimated at a minimum of £15,000 over its lifetime by the PDSA) purely to save their fellow citizens a few hundred pounds.[40]

VETERINARY COSTS ARE not the only downside of pet keeping. The medical profession's newfound enthusiasm for the practice contrasts starkly with the warnings it once issued about the health hazards of getting too close to animals. While improvements in routine veter-inary care have undoubtedly reduced these concerns, even compan-ion animals still pose some risks. Many of the infections we acquire from our pets are low-grade (see box), rendering their health impact over our entire lifespans difficult to estimate.

Common Health Risks Associated with Pet Dogs and Cats

Some dogs, puppies especially, still harbour the roundworm par-asite *Toxocara canis*, despite the ready availability of dewormers, and excrete the eggs in their faeces. Children occasionally become infected via contaminated garden soil. While the worms cannot de-velop fully in human tissue, they can cause damage when they start to migrate around the human body. This causes inflammation, usu-ally in the liver or lungs, leading to fever, and if the worms reach the eyes, blindness can result. Although the condition is not as common as it was, there are still around seven hundred cases of ocular larva migrans in the United States each year. In some parts of the world, particularly Asia and Africa, dogs are the main transmitters of rabies (usually through bites), but in the United States this dubious honour nowadays falls to bats, thanks mainly to the $300 million annually that dog owners spend on rabies shots.

The most serious cat parasite, in terms of risk to owners, is a single-celled protozoan, *Toxoplasma*. Unlike the roundworm, this or-ganism will happily live and multiply in any mammal (although it can only reproduce sexually in cats), and hunting cats are highly likely to pick it up from their prey. The protozoan, intent upon travelling from one body to another, performs the neat trick of making the rodents it infects easier to catch by reprogramming some of the nerves in their brains to decrease their fear of predators. This may have given rise to the fable that toxoplasmosis transmitted through cats can cause schizophrenia. Worldwide, about half the human population is or has at some time been infected by *Toxoplasma*, mostly with few ill effects except in people with extremely fragile immune systems. It does pose a risk to the human foetus as it can cause birth defects; thus, in the past it was standard medical advice for pregnant women to rehome their cats, or at least to refrain from cleaning out litter trays, since, as with *Toxocara*, *Toxoplasma* cysts are transmitted in faeces. However, there are many other sources, including undercooked meat and unwashed vegetables or fruit. Nowadays cat ownership is not especially associated with the infection, unless one is a cat breeder.[41]

More generally infectious than toxoplasmosis is cat scratch fever, caused by a bacterium spread by fleas (most fleas on cats are actually dog fleas, the cat flea being rather rare). After a delay of a week or so, the site of the scratch turns red, and the lymph glands on that side of the body swell up; complications are rare but include an often fatal inflammation of the heart valves.

Over our long coexistence with cats and dogs we have probably co-evolved to cope with most serious diseases that we might transmit to each other. And yet allergies to both dogs and cats have become more common, as a component of the epidemic of asthma, eczema and hay fever that has recently engulfed the Western world. That these disorders run in families and have a genetic basis does not alone explain the sudden increase in the number of people affected. Rather, changes in lifestyles during the second half of the twentieth century are the likely culprit. Some scientists point to the widespread adoption in everyday life during that period of potentially irritating chemicals (preservatives, surfactants, fragrances), pharmaceuticals, and dietary changes (processed foods and formula milk for infants).

Others prefer the 'hygiene hypothesis', which holds that, due to our urban lifestyles and hygienic houses we encounter far fewer allergens in infancy than we used to and hence do not 'train' our immune systems properly. Consistent with this theory is the lower incidence of these disorders in farming families, whose children get exposed to a wide variety of livestock allergens and associated micro-organisms from an early age.[42]

Remarkably, the medical profession remains divided as to whether childhood exposure to pets promotes susceptibility to or protects against allergies. Despite both species' producing their main allergens on the skin, exposure to dogs may have different effects from exposure to cats. Having a dog at home seems, if anything, to protect children against developing a dog allergy, whereas the reverse may be the case for cats and birds – although many studies have found no association at all. There seems to be a complex interplay between the genetics of the child and exposure to cats: in susceptible families, cats increase the risk of wheezing; in families with no history of allergies, they may protect against asthma. Living with a pet may not be essential for exposure, anyway. In places where pet keeping is common, even children who don't have pets at home encounter both cat and dog allergens at nursery or preschool, as children with pets carry animal hair and dander on their clothes.[43]

Perhaps most remarkably, from the perspective of understanding the bond between owner and pet, people often resist allowing an allergy, unless it is particularly severe, to dissuade them from owning animals. In one study, 80 per cent refused to give up their pets on medical advice, and 70 per cent acquired new pets after the original one had died. A medical team once consulted me about their difficulty in persuading pregnant women to part with their cats, even when their medical histories indicated that their unborn children were likely to become allergic. Surprisingly, these women were no more attached to their cats than was typical for their age and reproductive status; evidently they deemed their bond with their cat more important than medical advice, however well intentioned. Some parents refrain from obtaining pets when a child has a hay fever–type reaction to them, but pet keeping is at least as common in families prone to allergies as in those not susceptible. (Owners who surrender their animals to rehoming charities often cite allergies as their reason; such claims, however, rarely get checked, and some charities believe that allergy simply serves as a plausible smokescreen for some other, less socially acceptable motive.)[44]

OVERALL, FURTHER RESEARCH has not borne out the promise of early studies. Owning a pet doesn't guarantee future health, physical or mental. It may confer subtle benefits that depend on some combination of the pet and the owner's personality and lifestyle, but scientific (though not popular) support is dwindling for the notion of the pet as panacea for the well-being of adults.

The good-news stories that have stood up to scrutiny are few and far between. Some autistic children seem to benefit enormously from the company of the right pet (but others do not, recoiling from animals to almost the same extent that they do from humans). Stroking a dog or cat is pleasant but doesn't seem to produce any lasting health benefits. Walking the dog probably needs to become jogging the dog to have any effect on longevity.

One finding has proved difficult to debunk: that the company of a dog (and possibly other animals) bestows an aura of trustworthiness. This 'social bridge' effect not only accounts for why dog owners tend to make friends with other dog owners but may also underlie much of the success of animal-assisted therapy: the therapeutic effect may come primarily from the human handler, with the dog or other animal inducing a positive attitude in the patient. Science may have been slow to recognize this effect, but the advertising industry, which often includes animals in advertisements to enhance the image of the humans represented, seems to have been aware of it for some considerable time. (As presidential candidates, Franklin Roosevelt, Richard Nixon and Ronald Reagan took care to be photographed with their dogs (see nearby illustration). The recent spate of cats in Downing Street, ostensibly brought in to control mice, may have distracted from some of the more unpopular decisions of the Conservative administration.)

I have found the most interesting feature of the 'pets as panacea' meme to be the enthusiasm with which the general public has accepted it. Those who promote this idea, undoubtedly from genuine commitment, have found an audience eager to accept any good news to counteract the occasional spikes of bad press directed at dogs – especially 'dangerous dogs'. What about pets today causes many people to accept as unvarnished truth notions that their parents would have found deeply implausible? Why are we so readily suggestible when it comes to our animal companions? What is this deep-rooted need to connect with animals that many people seem to feel so powerfully? Perhaps, like much else in human nature, all of this begins in childhood.

President Roosevelt with his Scottish terrier Fala.

CHAPTER 4

'We Got the Dog for the Kids'

KIDS LOVE ANIMALS. They are the central characters in about a third of children's books. The percentage is even higher in books written for preschoolers (42 per cent of the books my wife and I keep for our grandchildren feature animals as the lead characters, and animals play supporting roles in many more).[1] Words for animals – especially 'dog' and 'cat' – are among infants' first utterances (sometimes preceding 'dada'!). The human species' obsession with animals evidently begins at a young age – and for some may actually peak during childhood.

That this fascination starts so early provides strong evidence that the human mind is hardwired to attend to and gather information about animals. Biologically speaking, this is quite remarkable. Most animals, apart from parasites and some specialist predators, are primed to learn the social habits of only their own species. (Understanding the behaviour of one's rivals is essential to survival.) Human children – uniquely – incorporate animal characters into their close social circles, although nowadays these are more likely to be imaginary characters in books and on TV than real animals, which they do not yet have the skills to interact with. My two-year-old grandson Senan is currently fascinated by the TV series *Paw Patrol*, in which a pack of dogs carries out rescue missions (I suspect he's rather disappointed when real dogs seem more self-centred). The allure of animals for children suggests that something in the human mind reaches out not only to other humans but to other living beings as well – and this prospecting starts in the first year of life.

This phenomenon seems so obvious that it gets largely overlooked: social scientists take it for granted, focusing mainly on the roles that animals might play in childhood development and education. People widely assume that the right kind of contact with animals benefits children. To quote one American mother, 'Our pets bring out the best in the kids in responsibility, kindness, affection, first-aid, and concern for other living things.' Parents also seem to believe, without much evidence, that pets help prepare children for later life experiences, such as pregnancy, birth, the rearing of offspring and the death of a loved one. We make much of pets' apparent power to imbue a sense of empathy, compassion and concern for welfare. Many claim that exposure to pets boosts children's immune systems, compensating for our extreme levels of household hygiene (by historical standards).[2]

Based on their own childhood experiences with pets, parents may project the benefits of pet keeping onto their children. Even as adults, people who grew up with pets tend to look back on them through rose-tinted spectacles. Take the following, from an article in the professional journal *Family Process*: 'In my case, growing up as an only child with two working parents, my dog Rusty welcomed me home from school [if he was ever left alone, he probably suffered from separation distress], shared my milk and cookies [would he have shared with her, given the choice?], and curled up close *to help me with my homework* [my emphasis].' Again, such reminiscences probably tell us more about the fascination and enthusiasm that many people have for animals, especially their own pets, than about any universal benefit to children's development.

Pets have acquired a reputation for benefitting children's development in ways perhaps more apparent than real. As with pets' perceived health benefits, we must ask one question: Why are we so ready to imbue our animal companions with seemingly magical powers? It's natural for parents to instinctively exploit their children's obvious fascination with animals as a bridge between their respective imaginations, but that doesn't explain where this fascination comes from or why parents find it so natural to believe that their children will benefit from a relationship with an actual – as opposed to fictional – animal. Quite plausibly both are responding to instincts established during the evolution of the human race rather than some affectation permitted by modern affluence.

PARENTS IN THE WEST tend to believe that pets teach responsibility. Families thus tend to acquire new pets when their children are between five and twelve years old, irrespective of whether the family is large or small and already has other pets or not. Parents hoping that their children will become more conscientious as a consequence of caring for an animal will probably be disappointed, however. Several studies indicate that children's interactions with pets are more sociable than responsible; consider the title of one such paper: 'Mum Cleaned It and I Just Played with It'. A minority of school-age children who contribute to the care of pets also help around the house, significantly more than those without pets. But this in no way proves a causal relationship. Perhaps some parents simply expect more assistance from their children and that includes help with cleaning out the hamster cage as well as doing the dishes.[3]

Some argue that being around pets can also improve a child's cognitive skills. On the surface, this idea sounds plausible – we are increasingly aware that children do not acquire 'intelligence' only at school. For example, physically fit kids seem to have fitter brains, processing information more quickly than children who take little or no exercise (a conclusion backed up by brain imaging). Acquiring the complex coordination skills involved in a sport such as tennis has tangible developmental effects. Learning to play a musical instrument likewise has far-reaching benefits, including improved linguistic skills, attendance to tasks, working memory and cognitive flexibility. Unfortunately no direct evidence indicates that learning to look after a pet might have more general effects on a child's mind beyond acquisition of the skill itself, despite many parents' beliefs.[4]

PROPONENTS ARGUE THAT raising children in the presence of animals teaches empathy – not just for animals but for people as well. The theory holds that in the absence of empathy, violence can become the norm and that those who maltreat animals when children go on to mistreat their nearest and dearest when adults. Much has been made of the supposed connection between cruelty to animals and to people, perhaps not least because it enables animal and human welfare charities to make common cause with one another. People assume widely that children with wholesome, nurturing relationships with pets somehow develop a greater capacity for wholesome, nurturing relationships with people, and that this benefit will last into adulthood. This idea relies, however, on a somewhat simplistic concept of empathy, as if it were a skill, like riding a bicycle: once learned, never forgotten. Professional psychologists view empathy as a complex, multi-level process that involves many aspects of a person's genetics, personality and attitudes. We must take care in assessing which elements of 'empathy' having a pet in the home might affect positively.

Across mankind as a whole, some people seem much more empathetic than others, and we are coming to understand the factors behind these differences. Women tend to be more empathetic than men, and empathy definitely runs in families. Empathy levels relate to parental treatment during childhood and also entail a genetic component, particularly with regard to the more intuitive, 'gut-feeling' aspect of empathy, the version that makes us well up when observing someone in tears (less so the more conscious side of empathy that enables us to guess what others are feeling). An essential component of human social life, empathy underpins unselfish behaviour and inhibits aggression: people who lack one or other of its components are generally disadvantaged and may be labelled 'autistic' or 'psychopathic'.[5]

Most people are born with a certain degree of empathy, but the extent of its expression develops as we mature (hence the claim that contact with animals might 'grow' empathy, of which more later). Infants display seeming empathy from a very young age, well before they have a clear sense of self: on witnessing an older sibling fall over and hurt herself, a ten-month-old will immediately look distressed and may seek comfort in his mother. By eighteen months of age, as their sense of identity matures, many children will not only have come to distinguish between their own distress and that of others but may also attempt to comfort the child who is distressed or try to enlist help, for example from their own mothers. Most three-year-olds can realize that other children may react differently than they do to particular situations, and by school age many express themselves emotionally when told about, without actually witnessing, another person's feelings. Older children can learn to incorporate much more general concepts into their empathic reactions, for example, understanding that other children may not enjoy as happy a home life as they do.

Many consider this most fully developed form of empathy to be a foundation of social justice. The impulse to help those worse off than oneself – for example, the chronically ill or the poor – probably has an emotional basis. Such ideas form the roots of modern liberalism, as originally proposed by John Stuart Mill (1806–73) and his mentor, Jeremy Bentham (see box on p. 126).

The idea that empathy for animals relates to that for humans has been around for centuries. The German philosopher Immanuel Kant (1724–1804) thought that kindness to animals reinforces kindness and fairness towards people. William Wilberforce, leader of the movement to abolish slavery, also helped found, in 1824, the Royal Society for the Prevention of Cruelty to Animals (RSPCA), and John Colam, secretary of the RSPCA in the 1890s, was instrumental in setting up the London (now National) Society for the Prevention of Cruelty to Children, which initially shared the RSPCA's boardroom in Jermyn Street. Indeed, the very first child-protection charity in England, the Liverpool Society for the Prevention of Cruelty to Children, came into being at an RSPCA meeting called to discuss the building of a new home for dogs. In the nineteenth century, links between child and animal protection were equally strong in the United States: in 1908, over half of the 354 American anti-cruelty organizations were 'humane societies' combating abuse of both animals and children, with only an eighth restricting themselves exclusively to children. Theologian, physician, Nobel Peace Prize winner and vegetarian Albert Schweitzer propounded respect for all living things in his 'Reverence for Life' philosophy. Mohandas Gandhi, also a vegetarian, said in a 1931 speech, 'The greatness of a nation and its moral progress can be judged by the way in which its animals are treated. I hold that the more helpless a creature, the more entitled it is to protection by man from the cruelty of man.'[6]

Of course, by no means everyone who cares for animals is a pioneer for social justice. In 1930s Germany, Adolf Hitler and several of his fellow Nazis somehow managed to reconcile their practice of vegetarianism and concern for animal welfare with a highly selective approach to the well-being of their fellow human beings. Hitler himself was an avid dog lover throughout his life, first adopting a stray fox terrier he named Fuchsl; even during the final weeks of World War II, he risked his life on a daily basis to walk his last dog, a German shepherd called Blondi. (His affection for Blondi did not deter him from using her to test one of the cyanide suicide capsules he obtained to avoid surrendering to the Allies. It worked.).

The correlation between empathy for animals and that for humans isn't always clear. Many of those who support organizations with a strong stance on animal rights, such as Compassion in World Farming and the Vegan Society, express high levels of empathy for animals in general but lower-than-average empathy for humans. By contrast, people who work in animal shelters are generally highly empathetic towards both animals and humans. However, farmers, who arguably have as much actual contact with animals as shelter workers, express less empathy for animals than do members of the general public. Pet owners tend to show more empathy towards pets and parents more towards children.[7]

People arguably make particular choices in life – for example, working in animal rescue, or giving up meat, or adopting a pet – because they already feel an unusually high empathy for animals. If this is the case, where does such empathy come from? Does it stem from direct involvement with animals? Does it reflect the

Raising puppies makes people more empathetic towards dogs.

influence of childhood experience or formal education? One Swed-ish study suggests that living with a pet does actually enhance the owner's empathy for animals in general – although a feeling of empathy for animals may have spurred that individual to take in a pet in the first place. Specifically, the study found that people who had raised puppies showed particularly high empathy towards pets, as if the act of rearing a young animal itself enhanced their feelings.[8]

All this variation undermines the idea that empathy, for both humans and animals, is some kind of universal currency that some people have more of than others. Indeed, one study estimated that empathy for people overlaps only about 5 per cent with empathy for animals, for men and women alike. Such a weak connection would render the existence of 'universal empathy' highly unlikely.[9]

Almost all people experience empathy of some kind and degree, but the emotion gets triggered in several ways. The research indicates that empathy for animals and empathy for people are genuine but distinct categories, even though people do vary in precisely which animals or humans they express the greatest empathy for (usually those individuals, of whatever species, that they know well or feel most similar to). Therefore, animals and people may possibly evoke empathy through different brain pathways that have evolved separately and for different reasons. The extent of the activation of these pathways may depend on their initial strength (a genetic factor) and also – and almost certainly more importantly – on the experiences of people and animals that each individual has from early childhood onward. Providing children with pets isn't in any clear sense the best way to teach them empathy for people. The connections between the two are too weak.

DESPITE RESEARCH EVIDENCE to the contrary, many continue to believe that concern for animals goes hand in hand with concern for people and that as one grows, so does the other. Thus many parents and educators persist in maintaining that contact with animals has a civilizing influence on children. The website of the charity American Humane extols the benefits of a lesson plan for five- to seven-year-olds titled 'What a Dog Needs': 'In this lesson, students will learn about some of the things dogs need in order to be happy, healthy and comfortable. Children who experience such lessons may become more empathetic adults who care not only about animals, but also about each other and the environment.'[10]

None other than Boris Levinson, the founder of animal-assisted therapy, promoted the idea that pets are actually good for all children's well-being. In *Pet-Oriented Child Psychotherapy* he wrote, 'The pet, in a sense, becomes the mirror in which the child sees him(her) self wanted and loved not for what he (or she) should be or might be or might have been but for what he (or she) is.' Our current understanding of dogs' and cats' minds generally supports the latter part of this assertion: that pets likely have far less complicated notions and expectations of children than their parents do. However, whether children inevitably value the family pet for loving them unconditionally is less clear. Perhaps children interact with household animals simply from instinctive fascination rather than because they obtain some deep emotional reassurance from a pet's attention (presuming they receive it – not all cats get along with children!). Pets more likely become objects of strong attachment when other aspects of the child's emotional well-being are less than perfect – perhaps during times of illness or loneliness.

When asked, children usually include pets (if their family has any) on the list of the most significant individuals in their lives, but it's not entirely clear why. Perhaps they do so simply because the pets are familiar and always present. Some evidence, however, suggests that pets provide some children with a certain amount of social support: in one small study of families with pets, parents of five-year-olds not thought to use the animals as confidants rated their children as more anxious than did the parents of those who did seem to have close relationships with the family's pets. Several studies have shown a link between children's self-esteem and their involvement with animals, especially in early adolescence, but cause and effect are unclear. Might children who lack confidence simply prefer other ways of spending their time?[11]

Few studies directly address the question of whether caring for pets might make children more empathic, but two early investigations by Robert Poresky suggested that the two might at least be associated together, although the assumption is usually that it's the pets that are causing the increase in empathy. In the first, he presented midwestern children aged three to six with four situations: 'a child has just lost its best friend', 'a child is being chased by a big scary monster', 'a child really wants to go out but is not allowed', and 'a child is going to its most favourite park to play'. He then asked the children how they – and the hypothetical child – would feel in such a situation (the 'right' answers being sad, scared, angry, and happy respectively). He then presented the same children with the same situations but with 'dog' instead of 'child'. Poresky found that the twenty-six children who had a pet at home were no more or less empathetic towards other children than the twelve who didn't, but of the twenty-six, the stronger the empathy the children expressed towards the dog, the more empathy they displayed towards the child. In a second study, Poresky showed that the stronger the parents rated the bond between the child and the pet, the more empathetic the child appeared to be.[12]

Finding that children who care about their dogs are also more caring towards other children doesn't prove that one leads to the other. Their personalities – or parents' influence – may have primed them to be more attentive of and sympathetic towards all living things. The few studies that have tried to show that children who learn to look after animals in a classroom setting subsequently become more empathetic towards their peers have been hampered by the fact that dogs can't, on their own, teach anything, so it is difficult to disentangle what the animal's precise effects might be. The enthusiasm of the volunteers who bring the dogs into the classroom could easily confer more benefit to the children than the animals themselves, a similar effect to the way in which elderly people respond to visitors with animals.

Caring for pets may also correlate with emotions other than empathy. In one seminal study, ten-year-old children who reported having had heart-to-heart talks with their animals were more empathetic towards people than those who didn't – but they were also more competitive. The opposite was true of the seven-year-olds in the same study: those with the closest relationships with their pets seemed the least empathetic. These rather confusing results may sound less surprising when we remember that there is little connection among adults between empathy for animals and for fellow humans.[13]

Nevertheless, the idea that pets teach empathy seems to have stuck. Uncertainties over the precise psychological mechanisms underlying empathy get glossed over in the push for more 'humane education' in schools. This is not, of course, to say that we should not encourage children to interact with animals. There is good evidence that teaching children to respect animals alters their attitudes towards animals – which is valuable in its own right. But it is much less clear whether their empathy for animals spills over in any meaningful way into more empathetic attitudes towards people.

MUCH OF THE IMPETUS behind proving a connection between empathy for animals and pro-social behaviour towards people seems to stem from the obverse: the idea that a single set of motivations drives violence towards people and animals alike. Some academics and humane organizations go even further, proposing not only that abuse of animals and violence towards people are linked but, more specifically, that the former somehow heralds the latter. Over the centuries, this alleged connection has received sporadic attention from philosophers and writers of fiction. One well-known example is William Golding's 1954 novel Lord of the Flies, in which a group of boys stranded on an island commit acts of cruelty to animals prior to fighting with and then killing one another. Anthropologist Margaret Mead believed that society should take animal abuse by children seriously as a warning sign of violent tendencies and stamp on it hard before it escalates: 'A failure of punishment here . . . can be as fatal or possibly even more fatal, than too violent punishment. The temptation to try it again, try something bigger, kill a cat instead of a bird, wring a dog's neck instead of a canary's, can take possession of the child.'[14]

In the past twenty-five years, science has been drafted in to support what had hitherto been more belief than established principle. Authorities on child development have long portrayed cruelty towards animals – even 'games' once deemed an acceptable part of boyhood, such as pulling the wings off moribund flies – as desensitizing the protagonists to violence. According to this view, anyone who harms or kills animals in a 'cruel' way (ignoring the large cultural differences in definitions of 'cruelty') is or will become a danger to his or her fellow human beings. Dr Randall Lockwood, a psychologist who has written extensively on the link between animal abuse and human violence, asserts, 'Those who abuse animals for no obvious reason are budding psychopaths. They have no empathy and only see the world as what it's going to do for them.'[15] The animal-protection movement often portrays cruelty to animals as habitual, associated with aggression towards partners or children, and a component in the intergenerational cycle of family violence, whereby abused children grow into abusive adults.

It's not difficult to find anecdotes to support this idea. Mass murderer Jeffrey Dahmer allegedly cut off the heads of cats and dogs and then impaled them on sticks; 'Boston Strangler' Albert DeSalvo trapped dogs and cats in crates and then shot arrows through them; 'Son of Sam' David Berkowitz poisoned his mother's parakeet. Many of the recent spate of 'schoolyard shooters' (nine out of twenty-three between 1988 and 2012) had committed 'up close and personal' acts of violence towards household pets – although mostly not their own – such as burning, drowning, kicking or beating the animal to death.[16]

More systematic studies, however, have provided only intermittent support for these ideas. Violent criminals and psychiatric patients are indeed more likely than others to admit to having committed acts of cruelty to animals when they were children, but the key word here may well be 'admit': perhaps they are more willing to disclose acts of violence of all kinds, wishing to project a 'hard man' image.

TWO DISTINCT HYPOTHESES currently attempt to explain the connection between animal abuse and violence towards people. The 'graduation hypothesis' holds that people who abuse animals as children become desensitized to violence and go on to become aggressive towards people. Despite the popularity of this idea, the data doesn't really support it: animal abuse often *follows* violence towards people. The 'deviance generalization hypothesis' posits that animal abuse and aggression towards people are simply two types or styles of antisocial or deviant behaviour. Rather than one leading inevitably to the other, both express similar underlying causes (dysfunctional family life, poverty, lack of education, psychological need for control or to express masculinity, alcohol and drug abuse – the usual suspects, in other words). The choice of man or beast as target, in this view, is largely a function of circumstance and opportunity: adolescent boys, the most common perpetrators of cruelty towards animals, may simply find stray cats most accessible as marks – and, perhaps more importantly, unlikely to exact retribution.

Moreover, most boys probably only commit such acts once or twice in a lifetime, perhaps exploring their burgeoning masculinity through what some psychologists have referred to as 'dirty play'. Children as young as six or seven may engage in play based on fantasy violence – for example, the dismembering of toy soldiers or soft toys. As boys approach adolescence, many add a social dimension to such behaviour, as when in groups of peers without adult supervision they

Boys teasing a kitten; detail from *The Kitten Deceived*, 1816, by William Collins.

egg one another on to carry out pranks of varying levels of aggression, such as ringing doorbells and running away, throwing eggs at passing cars, or committing random acts of vandalism or cruelty, the targets of which may occasionally include animals (see nearby illustration). Taking part in such prohibited activities seems to achieve two objectives: first, gaining status within a group that becomes bound together by shared complicity in such acts; and second, exploring the individual's capacity to achieve control over situations by deliberately exceeding the limits of what society deems acceptable behaviour.

Such rites of passage are probably as old as humanity itself. The abuse of animals by male children was noted by the Greek philosopher Bion of Borysthenes (and subsequently quoted by the more famous Plutarch) over 2,000 years ago: 'Though boys throw stones at frogs in sport, yet the frogs do not die in sport but in earnest.' In William Shakespeare's King Lear the Earl of Gloucester's lines include, 'As flies to wanton boys, are we to the gods: they kill us for their sport.' Yet when outside their peer groups, the same boys may behave impeccably, and most grow up to become upstanding and law-abiding citizens. The cruelty they have displayed, though deplorable in itself, may represent a testing of boundaries that ultimately helps them grow into well-balanced adults. The sort of inner rage that will subsequently lead to adult criminality and prison may motivate some such acts but almost certainly not the majority.[17]

Such niceties were largely ignored when, in the early years of the twenty-first century, several animal-protection charities, notably the American Humane Association, forged an alliance with those involved in preventing maltreatment of children, echoing the origins of many of these organizations as dual-purpose bodies. In the United States these constitute the National Link Coalition, aka 'The Link', with the additional inclusion of organizations concerned with domestic violence and the abuse of the elderly. In the United Kingdom, the RSPCA, the National Society for the Prevention of Cruelty to Children (NSPCC), and other charities from both the animal- and the child-protection sectors formed the Links Group, which put together a pamphlet, 'Understanding the Links: Child

Abuse, Animal Abuse and Domestic Violence', highlighting the possibility that abuse of pets could form 'part of a constellation of family violence' and repeating the claim that childhood cruelty to animals may lead to violence towards people in adulthood. Despite criticism from some quarters, promotion of such ideas shows few signs of abating: activists in Scotland launched a scheme as recently as 2015 to encourage vets (as well as firefighters, dentists and hairdressers) to become proactive in the detection of domestic violence, based 'on an American model' (presumably, The Link).[18]

Research and case reports support many of the claims in these various documents. Dogs that have been abused are frequently aggressive enough to bite children in the household, perhaps because, as a result of maltreatment in their previous homes, they have become prone to biting as a first line of defence; more likely cruel treatment by their current owners has rendered them chronically fearful and likely to respond to any provocation with violence of their own. It is also well documented that the pets of abused women may be used as pawns in their struggle with (male) abusers, who may threaten to hurt or kill these animals if their victims do not comply with their wishes. More generally, the deviance generalization hypothesis predicts that any household in which there is adult abuse of pets will likely involve other violent exchanges, including abuse of children.

Any assertion of a direct correlation between childhood animal abuse and future domestic abuse is potentially dangerous. The statement in the NSPCC document that 'sustained childhood cruelty to animals has been linked to an increased likelihood of violent offending behaviour against humans in adulthood' rests only loosely on research. Its implicit conclusion that the first leads to the second is not only misleading but could have devastating consequences for anyone caught up in its over-enthusiastic application to individual cases. After combining a large number of studies carried out between 1979 and 2008, researchers concluded that somewhere between 35 and 40 per cent of non-violent ('normal') men had abused an animal at some point in their lives. The same percentages held for violent criminals (rapists, murderers and other aggressive offenders), contradicting the 'graduation hypothesis'. The better designed of those studies did indicate a difference, with about 25 per cent of violent criminals admitting to abusing animals, compared to about 15 per cent of nonviolent men. But deviance generalization can explain this difference just as well as 'graduation'.[19]

Are people convicted of cruelty to animals more likely to have committed, or to go on to commit, other types of crime? The few published studies on this subject have failed to establish any strong connection (in either direction or even overall). Even fewer studies have distinguished between 'sustained' cruelty and occasional acts of abuse committed during adolescence. Overall, little systematic evidence supports the idea that a child who is repeatedly cruel to animals will inevitably grow up to be violent towards people.

THE LINK DOCUMENTS also posit that children who are cruel to animals may well have been abused themselves: 'If a child is cruel to animals this may be an indicator that serious neglect and abuse have been inflicted on the child.' There is some underlying truth here: abused children are indeed more likely than others to abuse animals. However, if, as seems plausible, about one-third of teenage boys in the United States and the United Kingdom commit some kind of cruel or thoughtless act towards an animal, then we could take this statement to mean that child abuse or neglect is endemic and widespread in both countries, which seems highly implausible (and if true should deeply worry society as a whole, not just the agencies directly concerned with child protection).

This is one example of how misleading claims about 'the link' can be. The National Link Coalition in the United States portrays animal abuse as 'the tip of the iceberg: the way animals are regarded in a family is a window into interpersonal relationships and family dynamics. Investigators who find animal cruelty, abuse or neglect are rarely surprised to see other issues lurking beneath the surface.' Remarkably, for such a supposedly telling indicator, 'animal abuse' is rarely defined precisely. Cruelty connotes deliberate harm; neglect implies a more passive failure to attend to the animal's needs. Does the term 'abuse' cover both or only premeditated acts – the latter being more in tune with the connection to family violence? The hoarding of animals to the point where properly caring for them becomes impossible is a well-documented form of abuse, but no evidence suggests that it routinely co-occurs with violence directed at people. To take another example, many regard as cruel the use of devices that deliver electric shocks as a part of routine dog (or cat) training: if this widespread practice falls within the definition of 'abuse', does that mean that we should suspect every pet owner who owns a shock collar of abusing his spouse or children?[20]

Since violence within the home often proceeds for years before anyone notices, agencies' sharing of information and working together to unpick such connections on a case-by-case basis is wholly admirable. Yet, taken uncritically or over-enthusiastically, assumptions about a connection between animal 'abuse' (such as, say, neglect caused by temporary misfortune) and child abuse could potentially result in heavy-handed intervention. Encouragingly, little published evidence indicates that such connections are being made in practice, although this may be partly because UK courts keep most child-care proceedings private. One case, from 2009, which was reported when it reached the court of appeal after two years of wrangling through the justice system, involved a five-year-old girl 'torn away from her parents after an RSPCA raid' – although nothing in the report suggests that 'the link' prompted the initial removal of the child by social services.[21] Nevertheless, such sweeping generalizations linking very different categories of so-called abuse create fertile ground for police officers and child-protection agencies to jump to conclusions not borne out by the evidence.

DUE TO THEIR instinctive fascination with animals, children will readily pay attention in lessons about all kinds of creatures, from woodlice to elephants, often becoming more positive in their attitudes towards even those animals that they initially find creepy or disgusting. Teaching children how to care for animals in an empathetic way is thus, in itself, both worthy and effective. However, scientific study calls into doubt whether much of this rubs off on how those same children perceive other people. We must ask whether we might more effectively improve the latter by teaching children directly about human social ethics. Perhaps educators could use animals as initial examples to grab the children's attention, before subtly shifting to more general issues such as fairness and social justice (indeed, some versions of humane education do incorporate such an integrated approach).[22]

It also seems unlikely that humane education will have much influence on the prevention of animal abuse. The little evidence we have suggests that animal abusers, far from showing a psychopathic indifference to animal suffering, actually express a normal range of levels of empathy for animals. Many seem remarkably sensitive to the emotions of animals, apart from those that they have selected as targets for abuse. This casts doubt on whether empathy plays a crucial role in all animal abuse. At least one study suggests that the main predisposing factors for serious animal abuse may be prior sexual abuse, coupled with a capacity to construct violent fantasies (possibly prompted by that abuse), which subsequently get enacted using animals rather than people as the target.[23]

Scientific study has struggled to clarify the roles that animals play in the lives of children. Animals make a useful hook to engage children in learning about social justice, but involvement with pets, while enhancing empathy towards those individual animals and possibly animals in general, may have little or no effect on how children relate to each other or to adults. Quite a high proportion of children, mostly boys, do go through a brief phase in which they commit acts of cruelty against hapless animals, but almost all grow up to be responsible adults. Household pets may provide children with reassurance during times of misfortune, although such support is entirely imagined by the individual, whatever his or her age: pets do not have the cognitive ability to understand what they are providing under these circumstances. The phrase 'non-judgemental companionship', often applied to pets, differs entirely from the 'non-judgemental approach' adopted by health workers and counsellors: the latter is the deliberate and conscious suppression of outward signs of judgement, whereas pets are intrinsically incapable of making judgements.

As with the supposed health benefits of pets, the most interesting feature of beliefs about the animals' capacity to civilize children may be their persistence. Nowadays only a small proportion of children grow up to work directly with animals, but for much of human existence, both a facility with animals and an inbuilt curiosity about them would have been crucial to survival. This would have applied to hunter-gatherers going back as far as the dawn of *Homo sapiens*; even today such societies place great emphasis on acquiring and sharing detailed knowledge of the animals in the environment around them. Despite an inevitable shift in focus, the same capacities would probably have gained even more importance once our economies became dependent upon the successful husbandry of domesticated animals. From this perspective, parents who encourage their children to interact with animals are simply obeying an instinct that would have benefited their ancestors' survival, not because it civilized their children but because it enabled them to acquire crucial knowledge and skills in relation to the animals themselves.

Moreover, even very young children display an instinctive fascination with animals, not only in the soft toys they clutch but in their preference for fictional animal characters in books and on TV. Their subsequent dealings with pets may not make them more human in the sense of being kind towards people, but they do appear to make them more understanding of animals in general. Given that other animals have, until very recently, played such a large part in our own species' journey from ape to urban sophisticate, their appeal for children is not surprising. In helping youngsters to appreciate animals even nowadays, when instructing them in computer code or Mandarin might seem more immediately useful, we may encourage them to embrace an intrinsic, if no longer essential, component of human nature.

Jeremy Bentham – Father of Animal Welfare

Much of modern liberalism, as well as current thinking about animal welfare, we can trace back to one eccentric Englishman, Jeremy Bentham (1748–1832). Bentham appears to have been unempathetic himself – he avoided social engagements and, on the rare occasions when he did dine with friends, circulated a list of topics for conversation beforehand. His philosophy, however, was well ahead of its time. He considered that everyone had the right to happiness ('it is the greatest happiness of the greatest number that is the measure of right and wrong'), disputing the entrenched rights of the aristocracy, the established Church, the Crown, and the judicial system of the time, which he came to see as conspiring to defend their self-interests against those of the populace as a whole. He opposed slavery and capital punishment; he promoted freedom of the press, free trade, equal rights for women and the right to divorce, and (privately) supported the decriminalization of homosexuality.

Bentham was also an early supporter of animal rights at a time when utilitarian attitudes prevailed and cruelty to animals was rife, especially in cities (the RSPCA was formed in 1824 specifically to protect working horses). He delighted in the company of animals: at one time he shared his bed with a pet pig (whose name, sadly, doesn't survive) and he doted on a tom cat that he insisted on referring to as the Reverend John Langhorne. He also kept large numbers of pet mice – surely a temptation for the Reverend John. However, Bentham's opinions on animal welfare seem to have stemmed not from his direct experiences with animals but from the more general consideration that if animals could be maltreated because they could not plead their own case, the same applied to human infants and adults with mental illness and dementia. He enshrined this logic in perhaps his most famous quote: 'The question is not, Can they reason? nor, Can they talk? but, Can they suffer?'

Bentham graduated from Oxford University but, after his death, became more closely associated with University College London, where his body (apart from his head, which has been replaced by a wax replica), dressed in his original clothes, is preserved in a glass-fronted cabinet, from which it is periodically wheeled out, in fulfillment of his dying wishes. Although he most recently attended a university meeting in 1975, thirty years later he was present at a college dinner commemorating his protégé John Stuart Mill.

CHAPTER 5

One of the Family?

WHILE MANY OWNERS treat their pets as 'little people', these animals are also possessions. By definition, they belong to their owners, even though most retain a certain degree of autonomy. Like that shown towards dependent children, the care that owners devote to their animals evidently has a strong emotional basis; yet there are also differences, reflecting not only the species barrier – our pets can't reason with us, even though they have their own ways of answering back – but also their ambiguous status as part person, part property. Nevertheless, these are the closest, most affectionate relationships that most people have with any kind of animal. In dissecting these relationships, I aim to discover whether pet keeping simply relies on a careless blurring of the human and animal categories or reflects something unique about the way we conceive of our animal companions that stems from our own evolutionary past.

Despite the trend to imbue pets with 'rights' that approach those of human dependents, Western society distinguishes clearly between pets and children in one sense: how we obtain and dispose of them. The veterinary profession encourages euthanasia of pets that are suffering intolerably; euthanasia of children, even the terminally ill, is illegal in many countries and highly contentious everywhere. Likewise, while many animal-rehoming charities use the word 'adoption', the process of adopting a child is much more regulated than the transfer of an unwanted pet to a new owner. Far less stigma attaches to a person who surrenders a pet to a charity than accrues to a woman who produces a child she cannot raise and must therefore relinquish. Some pets may become 'little humans' in their doting owners' eyes, but the species boundary is still very real to society as a whole and (except possibly in the case of some animal hoarders) evident in the way that people treat their animals.

Stranded somewhere between child and chattel, pets occupy an apparent no-man's-land that can blur the reality of the relationship between them and their owners. So far, most attempts to characterize the emotional aspects of these bonds have, perhaps inevitably, used comparisons with relationships between people as their starting point. But are human relationships really the best analogy for the rich and diverse ways that people relate to their pets? Or is the human–animal bond sufficiently distinctive to demand a new conception?

OUR PETS UNDOUBTEDLY arouse strong emotions in us. When pop star Katy Perry divorced comedian-turned-actor Russell Brand in 2012, the celebrity couple's cats, rather than their financial fortunes, grabbed the headlines. There was no question that each should keep the cat he or she owned prior to the fourteen-month marriage, Kitty Purry and Morrissey. The dispute centred on who should get custody of Krusty, the kitten they'd adopted together in 2010. Unfortunately for Brand, he had already publicly undermined his claim by offering Krusty for sale on YouTube – if in jest and with his then wife's apparent cooperation. Hence Perry took Krusty – and promptly renamed him Monkey.[1]

Battles over who gets the family pets have become commonplace in divorce and separation proceedings the world over, highlighting the equivocal status of domestic animals. Should we view pets as possessions, to be divided up like ornaments or garden furniture, or as children, requiring negotiated and codified residence and contact arrangements? Whose interests should take precedence, the owner's or the pet's? If the former husband loves the dog more than his ex-wife does, but she will keep the family home, where should the dog live? Is it fair to the pet for the courts to divide its time between the former 'parents'?[2]

The shifting focus on pets' 'rights' is a recent phenomenon, though the notion that animals deserve special provision in divorce settlements is not. In tenth-century Wales, statute decreed that the husband could take one cat from the household, but all the others must stay with the wife. The cat's affectionate qualities, however, were not valued particularly highly at that time: cats officially commanded a price of four pence, but only if they could see, hear and kill mice, had all sixteen claws intact, and were known to be good mothers (there was no mention of any affinity for laps, and tomcats seem to have had no value whatsoever). By contrast, today's ideal cat is neutered and shows no inclination to kill anything.

In microcosm, such considerations highlight the ambiguous status of pets in today's society. Are they really 'part of the family', as many owners say, or do people just say so to describe an affectionate relationship for which no more precise term exists? If asked directly whether their pets are members of their families, typically around 80

After Hurricane Katrina, many Gulf Coast residents declined rescue because the authorities refused to allow them to bring their pets.

to 90 per cent of owners say yes, although somewhat fewer men than women and fewer rural individuals than urban. When misfortune strikes, many pet owners do their utmost to avoid separation from their pets, even possibly jeopardizing their own lives as a consequence (see nearby illustration).[3]

If completely unprompted, far fewer people include pets in their 'inner circle' of close family. In one study of family structures carried out in South Wales, only about 8 per cent of pet owners in the sample cited their animals (mostly dogs) in spontaneous descriptions of their personal social networks. Only about half mentioned their pets at all when interviewed in depth about whom they would turn to for support. Moreover, owners who did talk about their pets often asked the interviewer's permission before doing so, as if unsure whether they should include these companions in any serious discussion of their family lives. Many seemed to think that others might regard emotional closeness with animals as trivial, perhaps shameful, and even more awkward to put into words than the intimacy of the owners' relationships with 'real' family members.[4]

Such reticence may stem in part from conflicting notions of what constitutes a 'family' and who counts as 'kin'. The more formal definition rests on genetic relatedness and legal contracts such as marriage and adoption; yet many people's working definition of family includes those they cohabit with or rely on for support and perhaps those who rely similarly on them: hence they may exclude an estranged brother (50 per cent related and therefore 'kin' in the biological sense) but not hesitate to include a neighbour (genetically unrelated) who takes care of the family's children after school while their parents are still at work.

One factor confounding whether people identify pets as family may be that in Western society as a whole, owners' perceptions of their pets have changed quite significantly within living memory. Many older people and those of any age from rural backgrounds tend to value animals more for their usefulness than for any companionship they might also provide. Dogs were traditionally kept for herding, guarding and hunting; cats were mousers first and household members second.

These days, a minority of dogs and a few cats still fulfil these functions. Some farmers still maintain a colony of cats to keep rats and mice at bay rather than relying on chemical poisons. Hunting, pointing and retrieving dog breeds remain integral to today's field sports, and a plethora of new roles have emerged over the past century or so that make use of dogs' brains, brawn or sensitive noses: guide dogs for blind people, wheelchair assistance dogs, and dogs that can warn their diabetic owners when a life-threatening fall in blood sugar is about to occur. The attachment felt by the person benefiting (who may not be the legal 'owner') may be especially intense, reinforced by the daily assistance the dog provides, but we must nevertheless regard the relationship as more complex than just that of companion.[5] In addition, many dogs work within an institutional framework, such as those that check your bags at the airport as you rush to catch a flight; many of these animals live in kennels and have no single 'owner', although some do live in their handlers' homes when off duty and hence legitimately belong in the category of 'working pets'.

Owners of some types of pet may feel ambivalent about including them in their definition of family because they do not keep them primarily for companionship. If kept singly, cage birds – such as the parrot that converses with its owner or the canary that charms with its song and takes a daily flight around the room – may qualify as family members. Others appeal less as pets than as wildlife, albeit in microcosm: people generally value aviaries of finches or tanks brimming with tropical fish for their aesthetic qualities (even if owners can identify the individual animals). A few such collections may be little more than decoration, a more care-intensive version of the exotic houseplant or an expensive kinetic sculpture.

It's easy to presume that a dog or a cat is a pet first and foremost, but there are notable exceptions. Most breeders of pedigree dogs or cats will tell you that they love their animals; yet for some, showing rather than breeding is the driver. Pedigree animal shows are highly competitive, and anyone who attends such an event will immediately sense that the breeders and handlers are simultaneously members of an exclusive club and intense rivals for the prizes on offer. Once established as part of the showing fraternity, breeders can also enjoy the social network membership affords: in the words of one couple, 'We don't smoke, we don't drink. . . . [I]t's a hobby for us and it gets us out. . . . And we've got friends everywhere then, you see, all over the country.'[6]

Dogs especially can become extensions of their owner's ego. Breeders absorbed by the world of the show ring can often view dogs and cats as more than pets or a way of connecting with like-minded folk. They can become all-important to the owner's self-esteem, an essential source of pride and self-affirmation. Even people who would never enter a pet in a competitive show may select its breed to en-hance their self-image, though this motivation may fade in im-portance as the relationship with the animal blossoms.

AT ITS CORE, an owner's bond with a pet is emotional. The rela-tionship might have other facets, but in the absence of affection, the animal is a 'pet' in name alone. The key to understanding the owner-to-pet bond is to comprehend what that affection entails and how it compares with the affection that flows between two people.

In examining owners' feelings about their animal companions, most psychologists have homed in on the concept of attachment, which has had something of a convoluted history in psychologi-cal circles. According to most accounts, psychologist John Bowlby originally coined the term 'attachment' in the 1950s to describe his then revolutionary theory of the affectionate bond between a human infant and its mother, replacing the notion that this rela-tionship rested on the mother's feeding of the baby with the assertion of infants' fundamentally emotional attachment to their mothers. In this original formulation attachment should transfer better to the way pets feel about their owners than vice versa. (Indeed, some ethologists have attempted to borrow the concept to look at how closely pet animals bond with their owners, apparently achieving counterintuitive results – see nearby box.) Pets, like human babies, depend physically on those who look after them and are therefore inclined to stay close – or attached – to them.[7] An analogous situation among owners would entail complete emotional dependence on a pet and a self-perception as a sort of pseudo-parent. Such a scenario might plausibly describe a few extreme cases, perhaps involving individuals who obsessively hoard animals or have a psychotic relationship with a single animal, but does not pertain to the vast majority of pet owners.

The version of attachment that most closely approximates the owner's relationship to the pet emerged in the 1980s to describe a

---

Are We Part of *Their* Families?

Dogs and cats have very different ideas of what 'family' means because their wild ancestors lived very different lives. But both, in their own way, fit us into their family structures.

Before domestication, cats were solitary animals, with familial ties occurring only between mothers and their kittens and lasting only until the kittens were old enough to fend for themselves. Wild cats are solitary and, in our terms, self-centred. Science has no definitive answer to the question of how cats perceive us, but there are a few clues. Our cats greet us by raising their tails and rubbing around our legs – precisely what they do when meeting another cat they know well or consider a family member. The smaller of the two cats usually initiates this greeting, which probably explains why they go on doing it to us when most of us clearly don't know how we should respond!

Dogs are descended from wolves, which in the wild have extremely powerful family ties. Most notably, juvenile wolves often delay forming their own packs, instead staying with their parents to help them raise the next one or two litters of cubs. Although dogs, when left to their own devices, do form packs, they seem to have lost the urge to help their parents, instead choosing to compete with them once they are old enough to start breeding. One explanation for this change holds that domestication has repurposed the wolf's natural instinct to help the family to direct dogs to regard us as their closest 'family' – accounting for their legendary loyalty.

mother's feelings towards her child (that is, the reverse of Bowlby's original conception) as well as a key component of romantic relationships. Because adults have greater cognitive powers, the attachment they experience possesses a different flavour from that experienced by infants. Attachment prompts parents and partners alike to be proactive in their relationships, rather than reactive, as babies are.

When adults form attachments to children or to each other, some type of caregiving usually follows. Psychologists conceive of caregiving as a behavioural system separate from but complementary to attachment. It is most obviously evident in mothers of infants. Caregiving (the emotional part rather than the consequent actions, such as feeding or comforting) theoretically gets activated when the child is distressed or in danger, whether the cue is hearing the baby cry from its cot or seeing it crawl unheedingly towards the top of the stairs. Resolution of the situation deactivates caregiving, the subjective experience being a reduction in stress and increased contentment.

The relationship between owner and pet is therefore something of a hybrid between different psychological models of attachment. By and large, owners take care of their pets rather than the other way around – and the obvious reversals, such as guide dogs assisting blind people, are only partial and far from spontaneous, since they require a great deal of intervention (mostly in the form of training) from other people. Thus the mother-to-infant model is a better fit than Bowlby's original infant-to-mother model. On the other hand, some owners do express an emotional dependence on their pets that bears some resemblance to that characteristic of romantic relationships and those between grown-up children and their parents. The strength of the affection felt appears to be similar to that felt towards a parent and perhaps even more universal: in one survey, 94 per cent of dog owners said they felt 'close' to their dogs, as compared to 87 per cent who felt close to their mothers and 74 per cent who felt close to their fathers. Even cats comprehensively beat dads, at 84 per cent.[8]

According to one widely accepted model of attachment and caregiving, people's behaviour towards their attachment figures reflects how much emotional comfort they say they obtain from the relationship and how likely they are to turn to these people in moments of distress. Typically, only romantic partners outrank pets in their attachment value. The caregiving aspects of the relationship are better characterized by the extent to which owners attempt to stay close to, and especially in physical contact with, the objects of their affection, how they react when their pet becomes distressed, and how much they miss it when it's not around. Put simply, attachment stems from dependency, while caregiving reflects owners' belief that their pets depend on them.[9]

The caregiving part seems obvious: not only do owners attend to their pets' physical needs, but most value their pets for the physical contact they allow. When it comes to touching, pets often represent a much more accessible target than another person, because the contact is not emotionally loaded. Dogs and most cats accept a stroke or a pat without reservation, whereas physical contact between people, especially between adults, often raises questions about the initiator's motivations and is thus rarely so spontaneous.

Touch is so automatic with pets that it can be almost subconscious. When dogs push their heads between our knees, we can find ourselves tickling them behind their ears almost without realizing what we are doing. It's hard to resist at least trailing our fingers across a cat's tail when she rubs up against our legs. Even budgerigars like a tickle under the chin during their out-of-the-cage time (see nearby illustration).[10]

Pet owners obviously care for their pets in the sense of feeding them and attending to them when they're ill or injured, but caregiving is also an emotional experience. The $1 billion pet-treat market tangibly manifests owners' pleasure in providing their animals with enjoyable and interactive eating experiences, over and above sufficient nutrition for their daily needs. Furthermore, pet owners experience genuine distress when their treasured companion is in pain and relief when they (and their vet) have resolved the issue.

Owners undoubtedly vary in how much they miss their pets when away from them – for example, at work. Some opt to have a cat because they prefer to restrict their emotional involvement to times when they can be home, confident that the cat will cope with being

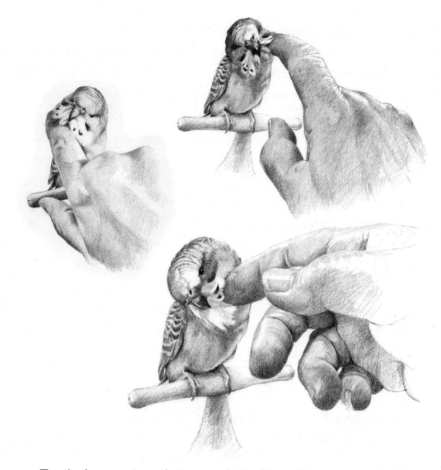

Touch plays a major role in our relationships with pets – even birds.

alone. Others choose a dog because they like the unreserved attention they get from their companion, perhaps not appreciating the
downside of the animal's emotional dependence. Most dogs become
upset when their owner disappears – despite popular opinion, the
company of another dog does not alleviate that distress. Cameras
that monitor pets and relay pictures to owners' smartphones are becoming increasingly popular – some even double as treat dispensers –
thereby satisfying owners' craving to see their pets. Dog owners may
be upset to see how agitated their dogs are in their absence, but this
will hopefully alert them to the fact that their dogs are missing them
even more than vice versa.[11]

IF THERE EXISTS such a thing as attachment to pets, in the original sense of the attachment that human infants feel for their mothers, it must stem from emotional dependence. To be an attachment figure, the pet must represent more than simply something to care for; it must also give emotional comfort and reassurance. For infants, the mother's presence provides at least two distinct psychological benefits: reassurance when the baby is distressed (she is a so-called safe haven) and encouragement in times of challenge (babies are much more likely to explore their environments from the secure base that their mother provides). Similar (but reciprocal) benefits obviously apply to well-functioning couples. Each partner comforts the other when needed, and the other's presence gives each more confidence in mildly stressful situations, such as demanding social occasions. Some evidence supports the idea that the company of a dog makes its owner more confident in mildly stressful situations (few cats have been tested, but on the surface they seem less promising candidates), so some kinds of pet do seem to meet the emotional aspects of the 'secure base' criterion.[12]

More broadly, however, abundant evidence indicates that at least some owners do rely on their pets for emotional support. Dogs, being generally more responsive than other species, seem to fit the bill best. As expected, in one study, adult owners from the United States and Canada rated their dogs more highly for the emotional aspects of caregiving than they did for the two facets of attachment, reassurance and encouragement. Nevertheless, over a third gave their dogs maximum ratings for both reassurance and encouragement. They seem to have interpreted 'encouragement' somewhat imaginatively as it includes such nebulous ideas as 'I can count on my dog to be there for me.' Such a statement might genuinely apply to working dogs, if an owner relies on the animal for his or her livelihood or to carry out day-to-day tasks. But owners of pet dogs may simply be referring to the reassurance that their dogs provide them in times of stress.[13]

Turning to reassurance itself, dogs often get rated more highly than other family members. Owners in the above-mentioned study described their dogs as more reassuring than their mothers, fathers, brothers, sisters, best friends and children. Only romantic partners beat dogs on this front – and very narrowly. A tendency for avid dog lovers to volunteer for the study may have biased the results slightly, but dog lovers certainly assert that they get a great deal of emotional comfort from their pets, especially in retrospect after the animal has passed away. As one former dog owner put it, 'My dog was the only psychologist I trusted; after his death I have nobody to tell about my problems.'[14]

Do cat owners find their cats emotionally reassuring? On the face of it, cats might be less suitable candidates than dogs because they tend to choose when and how to interact with their owners, whereas most dogs supply affection 24/7. While I have never in my own research made (possibly invidious!) comparisons between cats and romantic partners as effective confidantes, cats actually don't fare badly. Typically, I've found, cat owners say that they would possibly or even probably turn to their cats in nine common stressful situations (the full list I use includes feeling lonely, having a 'bad day', having trouble within the family, experiencing a temporary loss of confidence, feeling unwell, feeling depressed, experiencing problems with people outside the family, feeling nervous and losing a loved one). Incidentally, I quickly discovered that owners distinguish clearly between situations in which they would turn to a pet as a confidante and those for which the very idea would be ridiculous – for example, during times of financial stress. This is an entirely logical division, given that cats' paws cannot work a calculator, but in making such a distinction respondents showed that they did genuinely feel it worthwhile to confide in their cats (or dogs) when their problems were more emotional than material.[15]

This perception of reassurance requires some explanation, particularly when attributed to a cat. Cats and dogs have rarely been studied separately in this regard, but in one survey respondents rated cats as equal to the human members of the family in the reliability of their friendship; dogs scored slightly higher than the humans, as they did for companionship, on which cats performed less well. (Just to show that an overwhelming tide of sentiment for animals did not drive the data collected in this study, neither dogs nor cats performed as well as humans for the sharing of confidences.)[16]

Pet owners talk about the 'unconditional love' provided by their pet. As Sigmund Freud once observed, 'Dogs love their friends . . . quite unlike people, who are incapable of pure love and always have to mix love and hate.' Pets, lacking language, cannot publicly pass judgement, answer back, divulge confidences, or be influenced by what others say about their owners (in the study referred to above, ratings for conflict were highest for humans, intermediate for cats, and lowest for dogs). Thus the owner may freely project, within limits, any interpretation onto a pet's behaviour at any given moment. Cats, having somewhat inexpressive faces, give little away, often seeming either oblivious to or even dismissive of (as if they cared!) their owner's ramblings. Dogs are almost obsessively attentive to their owners, so are always ready and waiting to absorb a confidence.

THE STRENGTH OF the bond becomes most evident when a pet dies. Given that dogs and cats may live for somewhere between ten and twenty years, most owners inevitably go through such an experience at least once. Rationally, they know the pet is not immortal and that they will likely outlive it; nonetheless, most owners find themselves deeply upset when their animal companions die. Not all experience full-blown grief – studies estimate that between one in five and one in three may be profoundly affected – but for those who do, the emotional scars may not heal for several months. Symptoms can include a deep sense of sadness, uncontrollable weeping, and even a temporary subclinical depression. In the words of one 32-year old Israeli lawyer after the death of her dog, 'Before he died, I was so full of energy. My friends were amazed how many different things I was able to accomplish in a day. And now I'm exhausted and can't even bring myself to pick up my son from nursery school. So I leave him there, and his teacher takes him to her home to stay with her for a few nights.' Owners may take deliberate steps to rekindle memories of the relationship: 'In the evening I take one of my dead dog's blankets to my bed and I sleep together with him. In the morning, I kiss his leash and bless him.'[17]

Since all religions have much to say about death and bereavement, we might expect religious people to experience the loss of a pet differently from those with a more pragmatic outlook on life and death. Unsurprisingly, given their view that animals have no souls, most of the Abrahamic religions do not address the death of animals directly: for example, Judaism has no animal equivalent to the traditional seven-day period of mourning (shivah) that follows the death of a close relative. The *Catholic Herald* dismissed as a misunderstanding a *New York Times* report that Pope Francis had told a little boy whose dog had just died that 'Paradise is open to all God's creatures'. By contrast, Hinduism and Buddhism profess a continuity between human and animal minds rather than the dualism of Christianity. Advice on coping with pet loss occasionally appears in newspapers in India, and the Buddhist Phowa ceremony, the 'transference of consciousness at the time of death', can and has been used for pets. In Japan, pet funerals have become commonplace, even though the tradition of pet keeping by the Japanese middle classes is little more than half a century old (see nearby illustration). Funerals have even been held for defunct robot dogs (AIBOs), rendered unhealable by Sony's decision to stop manufacturing replacement parts and close its 'AIBO clinic' in 2014.[18]

Far from helping with the pain, religiosity (at least for Christians – no other faiths have been studied in depth) seems to make grief for a pet harder to bear. People who tend to feel anxious, angry or guilty (the personality type classified as 'neurotic') experience grief more deeply when a close relative dies; not surprisingly, the loss of a pet affects such people more than most. However, the only study so far reported found religious belief to be twice as powerful as neuroticism in predicting how intensely owners grieved. Specifically, those owners who wondered whether God had abandoned them when their pet died or even regarded the death as some kind of punishment were affected the most (the same link appears with the death of a relative). More surprisingly, intense grief for a pet

In Tokyo, grieving owners can rent a cubbyhole
at one of seven pet cemeteries, where they display
objects that remind them of their treasured pet.

also correlated – though less strongly – with attempts to seek solace
in religious belief. This is logical enough – a religious person feeling
stressed will likely try prayer as a solace – but also suggests that such
attempts are not particularly effective.[19]

Since Western religions, at least, seem to provide little consolation for bereaved pet owners, the recent mushrooming of 'pet-loss'
hotlines is unsurprising. The Blue Cross, a rehoming charity, offers a telephone service, as do the American Society for the Prevention of Cruelty to Animals and many of the US colleges of veterinary
medicine. A profusion of websites also offer advice. Since such services are confidential, we can't know whether the callers feel genuine
grief, as they might for a human relative; guilt, for example, over the
decision to request euthanasia; or mixed feelings about whether to
replace a recently deceased pet.

Despite the similarities between the loss of a pet and the passing of a close friend or relative, differences also colour the experience.

Nowadays in the West, few pets die of natural causes: most are put to sleep by vets. The decision to take the pet to the vet often rests with the owner, who may go through agonies of indecision when trying to strike the very difficult balance between the pet's deteriorating quality of life and its right to live as long as possible. Owners realize during this time that assessing the degree of an animal's suffering is more art than science, and few vets will make the judgement call for them. In the words of one owner, 'the only problem is – where do you draw the line – how long do you let a dog suffer? What is really critical for our decision to stop it?' Some take the cessation of eating as the critical point: 'Physically, the dog was already in a very bad condition, but as long as he ate – we continued.' Others imagine that the dog himself has made the decision: 'I saw in his eyes that he wants to die.' 'His eyes just said "Enough". 'The dog stood up and looked at me – and I understood that she was ready for it.'

Even if most owners can verbally justify their decision, they may still feel guilt. Did I wait too long to put down my pet? Or not long enough? The death of a beloved person may also invoke feelings of guilt, but these will likely differ from those felt when one has had to take a lead role in deciding the timing. Nevertheless, only about one in five owners reports serious feelings of guilt, and these may reflect a nagging feeling that they opted for euthanasia for their own convenience – for example, when the pet had become incontinent but otherwise appeared to be in reasonably good health.

Perhaps the death of a pet differs most fundamentally from the loss of a human family member in that many owners consider obtaining a replacement only a couple of weeks after the euthanasia. Not all, of course. At this point many are still grieving and unsurprisingly say such things as 'I'm afraid to get a dog because it hurts to bury him at the end.' Some may decide that no longer having to care for a pet, they can make a lifestyle change – for example, travel more. In some instances the replacement is almost immediate: 'I always get another dog a month after the death of the previous one.' Even those who grieve intensely often get over the death of a pet more quickly than they would that of a close family member: 'with a person like my mother, it was more of a long-term thing. But with a dog, it was very intense for a short period and then you've got over it. But with people then, it's not as intense, but it's for a longer period.'[20]

Even if the emotional impact is similar in the immediate aftermath of the sad event, the death of a pet is not the same as that of a relative or close human friend. This distinction points to a qualitative difference in the attachment between owner and pet – often intense while the pet is alive, it fades for many once the animal is no longer physically present. We can't really chalk this up to a failure of human memory. After forty-five years of pet ownership, I can recall the characters of each of my former animal companions no less vividly than I can recall my grandparents and parents (and there are photographs of both on the walls of my home). But like most other lifelong pet owners, I had no problem contemplating obtaining a new cat or dog a few months after one or other had died. In general, the attachment people feel for their pets may be intense, but fades more quickly than the bond with a human relative after his or her decease. In the long term, our brains treat the two differently.

PSYCHOLOGISTS WHO HAVE looked specifically for them have found a few similarities between emotional closeness to pets and the propensity to form affectionate relationships with other people. Contrary to the lazy stereotype of women as more caring, personality seems to play more of a role than gender when it comes to pets. As in other aspects of their emotional lives, women tend to express more affection for their pets than men do, but that may simply be because women tend to be more open about their emotions. In one study of mainly Caucasian American female owners – so not necessarily representative of all owners – those who scored high on the personality traits of conscientiousness (efficient and organized rather than easygoing or careless), agreeableness (friendly and compassionate rather than analytical and detached), and extraversion (outgoing and energetic as opposed to solitary and reserved) reported slightly closer relationships with their pets than others did. Conscientiousness, in

particular, may benefit the pet, which, after all, needs proper looking after. We know that people who score high on conscientiousness
have physically healthier spouses than average, so the same benefit
likely accrues to pets, though the research has not documented this
connection. In the same study, an anxious relationship with a pet
was (perhaps not surprisingly) more common in owners who already
had a disposition towards sensitivity or nervousness (this too may
carry benefits for the animal, since such people also tend to be risk
averse). Thus some personality traits seem to have similar influences
on relationships with pets as they do on relationships with people.[21]

Pet owners don't seem to form attachments to their pets using the
same rules as they do with people, however. For example, recent
studies have found that a willingness to share confidences with a pet
doesn't predict whether someone enjoys confiding in people, or vice
versa. Nevertheless, some people who prefer to remain somewhat
aloof in their human relationships do seem to feel a certain degree of
anxiety about using a pet as a safe confidant – perhaps they are unsure whether it is the right thing to do. Also, people with somewhat
anxious personalities may be a little uncertain about whether any
potential partner – animal or human – will love them back. Nevertheless, these links are fairly weak. Overall it appears that the way
most people approach and feel about their relationships with animals
does not correlate strongly with their corresponding view of people.[22]

This difference argues against the 'pet as child/friend substitute'
concept of pet keeping. People who take on pets apparently develop
their own way of relating to them, depending to some extent on their
default approach to relationships of all kinds but much more on their
individual experiences with that particular animal and others they
have owned in the past. Researchers have not studied this process in
detail, but for dog owners at least, the dog's behaviour can be much
more important in determining the quality of the owner–pet relationship than any effect of their own personalities. Whatever the details,
pet owners don't automatically apply the way they feel about relationships with people blindly to their animals.[23]

ATTACHMENT TO PETS also has an emotional downside. Pet ownership is not always a bed of roses: it's not all about attachment and caring and 'unconditional acceptance'. There has been little study of, and precious little comment about, what happens when a pet-owning family becomes dysfunctional. Nevertheless, my casual conversations with pet owners have often spontaneously led to discussions of people they know who have not gained as much benefit from their animal companions as they may have expected to – and these remarks are often couched in critical terms; for instance, the suggestion that these people 'should never have been allowed to have a pets'. Despite its long-lasting consequences, the decision to acquire a pet rarely gets taken half as seriously as decisions about which job to take, whether to get married, or when to have a child.

In some families, the addition of a pet can undoubtedly improve the balance of human relationships. Occasionally, the company of an animal will enable a deeply lonely person to go on living. A child who imagines that his parents do not love him may find solace in the affection of a dog or cat. Walking the dog can become a bridge to the outside world for someone who feels trapped in an unhappy home.

Pets can also negatively impact family relationships. Those who acquire pets sometimes fail to consider how their relatives might react, as in the case of a couple who had stopped visiting the wife's sister because 'they've now got a dog, and the dog doesn't like my husband. He doesn't like men at all!'[24] Children who are acutely allergic to dogs or cats can find themselves cut off from friends whose families start keeping pets.

Pet keeping is often a bumpy ride. When pressed for an answer, more than a third of cat and dog owners will admit to finding at least one aspect of their pet's behaviour problematic. For example, it seems likely that more pets regularly urinate or defecate in the house than their owners would like to admit. In a comprehensive survey of owners in a village near Liverpool, 5 per cent of subjects' dogs 'sometimes' or 'often' urinated indoors, and 4 per cent defecated. Many of the respondents had skipped this question, perhaps embarrassed to admit such failings to the interviewer standing on their doorstep. Many couples do not adequately prepare their cat or dog for the arrival of their first baby and worry that their newborn may not be safe with the dog that growls at it or the cat that tries to get into its cot. The majority of dogs, unless trained to cope, hate being separated from their owners and can show their distress by howling, barking, urinating or defecating, or destroying the furnishings. Some owners, unaware that the problem is soluble through training, give up going out in the evenings or at weekends – in extreme cases, one member of the family will decide to work from home or even give up work entirely. Cats that urinate in the home or dogs that fear strangers can become such an embarrassment that their owners stop inviting friends around. Such sacrifices are testament to the strength of the 'pet-as-family' concept: any decision to relinquish or have the pet put to sleep may be seen as a kind of 'Sophie's choice', the sacrifice of one family member to protect another.[25]

Owners who allow such a state of affairs to develop can be criticized for failing to manage their animals properly, and sometimes there is truth in this. Some believe the problems are the animal's fault and, once enough complaints have accumulated from within the family or from neighbours, will abandon the animal, soon to replace it with a younger and cuter model. Many, however, simply do not know that they can alter their pet's behaviour or have taken bad advice, tried, failed, and given up.[26]

Sometimes the addition of a pet can spell disaster for the family. Leonard Simon, a psychoanalyst practising in New York in the 1980s, interviewed hundreds of randomly selected pet owners. He said,

Not everything I heard was benign. With some people I became convinced that their lives would have gone altogether differently – and probably better – if there had been no pet. All too often I heard of wasted years and stagnant lives in which almost everything a person did revolved around his animal. I heard of divorces that might never have happened and I heard of some that probably should have happened long before and [after the pet died] they finally did. I heard of children that were neglected for the sake of a pet. I heard of children that might have been born if there had been no pet. I heard of children that were bitten by dogs that had given clear signs of serious jealousy but whose owners were unable to part with them.[27]

If devotion to a dog or a cat can sometimes place a strain on family relationships, owners of horses, it seems, commonly find that the time and money they must devote to their animal become grounds for conflict (see nearby box).

DESPITE CERTAIN SIMILARITIES, relationships with pets and with other family members also have differences – some subtle, others not. Owners receive constant reminders, however subliminal, that their pet is not a person: a 'no pets allowed' sign, their pet's engagement in its own 'private' species-typical grooming behaviour. A cat scratching its claws on the back of a chair, a dog seeking out and consuming a warm heap of horse droppings – these are other aspects of pets' lives that owners cannot and most likely do not wish to share.

Pets get treated sometimes like children and sometimes like possessions. This distinction is clearest at the end of a pet's life. Euthanasia is widely accepted as the best way of ending an animal's suffering but is controversial at best where the fate of another human is concerned. Moreover, the majority of owners eventually come around to the idea of replacing a pet that has died. No evidence suggests that the death of a pet leads to the separation or divorce of the couple who owned it, an outcome that can follow the death of a child.[28]

The bond that forms with a pet cat or dog seems more changeable than that between adult humans. Contrary to the 'love at first sight' and 'whirlwind romance' propounded by fiction and the popular press, psychologists conceive of lasting attachments between two people as taking several years to form; grief following the death of a beloved person can last a lifetime. Affection for a pet seems quicker to build (perhaps because of the pet's inability to contradict whatever ideas its owner has of it) and correspondingly quicker to dissipate – which is just as well considering the relative lifespans involved. Moreover,

Horse Owners – Members of *Their* Families?

Although few horses are pets in the same sense as dogs and cats, their relationships with their owners have some common features to those enjoyed by household pets. Most owners talk of the strong bonds of friendship that develop between them and their horses, and for many the bond also has physical elements, expressed through grooming and spontaneous displays of affection, such as hugging and kissing. Some say that their animals provide significant emotional support and relief from the stresses of everyday life. 'If I am stressed or unhappy, she has a lovely way of nuzzling me that makes me feel better!' However, unlike dog owners, most horse owners suffer little illusion that their horses love them or that they have to work to earn their trust. 'To my horse, I look like one big carrot.'

The most important component of the horse–owner relationship appears to be the high level of physical coordination that develops through riding, the sensation of being at one with the horse: 'we got to the point where I could just "think" a cue, and the horse would respond. It was like . . . one flesh, one heart, one mind. I think it takes years to come to this kind of understanding.' This high level of near-intuitive communication – to and fro – seems to lie at the heart of the almost spiritual closeness that many horse owners express. 'When I am with my horse I am closer to God than at any other time. When I am with my horse I am where I was meant to be, doing what I was meant to do. When I am with my horse I am truly complete' (though in my own experience the many companion horses that cannot be ridden, because they are unwell or too old, still receive undiminished affection from their owners, even if sometimes they disclose this a little wistfully).[29]

Overall, the horse–owner relationship appears to have three distinct dimensions. It does include elements of animal companionship, but unlike the relationship with household pets, it also entails participation in a high-risk sport. Moreover, many riders seem to imbue it with an almost spiritual dimension that may be not be unique but is often voiced more explicitly than is usual for indoor pets.

most owners' approach to their relationships with their pets – their 'attachment style' – bears little resemblance to the defaults they use when forming new connections with people.

There are some self-evident operational differences: you can talk to your cat about the problems you've been having with your neighbour, but the cat can't offer you any concrete advice in return, let alone pick up the phone and attempt to resolve the situation. The 'support' a pet provides is often largely imaginary (unless it also happens to double as a working animal). The differences in quality between our relationships with our human families and those with our pets may stem at least partly from their lack of language and consequent ability to act as an uncritical foil for our emotional needs – thus from their seeming 'unconditional love'.

It is also likely, however, that across the evolution of our species, more pragmatic considerations have affected the way we think about animals, even those we have taken into our homes and bonded with. A shepherd must understand every detail of his dog's character if the two are to have the intense rapport needed to outwit sheep. Yet over his working lifetime, a shepherd will own several such dogs and need to be able to form a new bond as soon as each successive dog becomes too old and infirm to work: he cannot let his grieving for one dog cloud his ability to work with the next. Surely it's not out of the question that the different ways we grieve for people and for animal companions have a basis in how our minds evolved.[30]

The question remains, Do we really think of our pets as members of our families, or is this just a convenient metaphor? Very few studies have probed this question effectively. Most have asked pet owners directly whether they think of their animals as part of their families, and the owners – perhaps somewhat defensively or at a loss for an alternative answer – have mostly agreed that they do. Indeed, the concept of the all-species-inclusive family is now so pervasive in the media that a researcher would be hard-pressed to question people about their pets without the family concept popping into their heads. Yet, despite this possible bias, there are enough similarities

in the ways we form relationships with people and with pets to indicate that they must have some underlying psychological processes in
common.[31]

Pet keeping is only one of humankind's many types of interactions
with non-human animals, and so pets must also trigger other ways
of thinking. During our own evolution, animals have preyed on us,
and we have been predators ourselves. We have exploited animals as
food, for transport, and as guards. We have an aesthetic relationship
with animals distinct from that which we have with humans. It may
be a convenient shorthand to talk of pets as family members, but the
fact that our relationships with them do not track those that we have
with people, together with the ease with which many of us replace
our animal companions, suggests otherwise. Pets occupy not so much
a no-man's-land between person and possession as a unique space of
their own in the complex tapestry that is human society.[32]

If pets are neither chattel nor child, our brain's perception and
interpretation of their behaviour should be equally distinctive. If pet
keeping is an intrinsic part of human nature, then it must rely on
thought processes that involve more than a confusion of the ways we
think about other people and the ways we think about animals. It's
logical that the owner–pet relationship should borrow psychological mechanisms from both, because evolution favours modifications of
what already exists rather than inventing new pathways; yet it should
also have unique features. Some of these emerge in the language that
we use to describe what our pets are thinking, which betrays our automatic lapse into anthropomorphism. We express others in our feelings about them, our reactions to their apparent cuteness, and our
enjoyment of their company.

CHAPTER 6

Imaginary Animals

Why Our Brains Search Relentlessly for Life

A s a ubiquitous feature of modern life that dates back to the dawn of our species, pet keeping must have roots in the workings of our brains. And yet we have been telling ourselves a different story. Over the past half-century or so, scientists have devoted much effort to examining instead the putative benefits of pet ownership – alleviating stress, extending our lifespans, healing those with disabilities, making our children kinder people, filling the gaps in our ever-shrinking families. Many of these benefits have proved largely imaginary, or at least far less substantial than promised.

So why, then, do many of us feel the urge to keep dogs and cats at all, now that they no longer serve their historical functions? And why are so many of us inclined to believe that these animals can have such profound effects on our well-being?

The answers to these questions must lie in biases intrinsic to our psychology, and as a biologist by training, I'm inclined to seek evolutionary explanations for them. We react to animals quite differently from how we respond to other features of our environment, and despite a great deal of overlap in the nature of our interactions with animals and with people, there are, as we've seen, differences, some obvious, others subtle. A conversation with a cat or a dog is far more one-sided than one with a person, which may indeed be part of its appeal: it allows the owner to imagine the 'unconditional support' that the pet supposedly provides.

Why do we project so much onto our pets? We have a strong tendency to invoke human thoughts and feelings in all kinds of situations – not just those involving animals – where there's no immediate evidence that another human mind is actually involved. This anthropomorphism is evidently a part of human nature, and although it probably did not evolve specifically so that we might share our lives with animals, nowadays it probably finds its most common expression in our relationships with pets.

Anthropomorphism appears to be a by-product of the uniquely human habit of guessing what other people are thinking. When we do so, we make the reasonable assumption that they have minds like ours. When trying to guess an animal's intentions, we tend to fall back on the same mechanisms, unless we know the animal well and can base our predictions on its behaviour in similar situations in the past. We anthropomorphize animals, and most of the time the process works well enough, even though we know that their worlds are not identical to our own and therefore their thoughts and motivations must differ in some ways from ours. In more superstitious times, we also imagined that an assortment of important yet (at that time) unpredictable entities had minds and intentions: the sea, the weather, the seasons. But today our pets are the chief target of our instinctive attribution of humanlike thought processes to anything and everything that appears to have a mind of its own.

IN AUGUST 2011, gangs rampaged through the streets of the Tottenham area of London, looting shops and setting fire to commercial buildings. Copycat riots instantly sprung up in other parts of the capital and also in major cities such as Birmingham, Nottingham, Bristol, Lincoln and Manchester. To this day, the reasons for the disturbances remain a point for debate, with the finger of blame alternately directed at racism, class envy, economic decline, a breakdown of social morality, and gang culture. Whatever the causes, five civilians died, and many were injured; over several days almost two hundred police officers, as well as five police dogs, were wounded during the struggle to contain the rioters.

One of those dogs was subsequently awarded the 'animal OBE' – technically speaking, an Order of Merit medal bestowed by the People's Dispensary for Sick Animals (PDSA), a charity that provides free veterinary care for low-income pet owners. Ten police horses also received identical medals. According to the *Mirror*, 'A police dog whose skull was fractured during the London riots has been given an "animal OBE". Linpol Luke, known as Obi, was hit by a brick as he braved missiles, petrol bombs and dense smoke . . . The German Shepherd was among a group of 10 [*sic*] animals who were presented with the PDSA Order of Merit for their heroic efforts as they braved the chaos to support police. Obi's handler Phil Wells said "He's one of the team – at the top of his game and a leader in his class with a finely tuned nose. But once home he's part of the family, a gentle giant and a cuddly bear."' Even cats occasionally receive recognition: in 2015 the Los Angeles Society for the Prevention of Cruelty to Animals handed its annual 'Hero Dog' award to Tara (a cat) who appeared (on video) to have stopped a dog from attacking a child innocently riding his bike in his front garden.[1]

Can animals really be heroes, or are their seeming heroics entirely a figment of our imagination? Citing animals for bravery evidently incorporates the kind of feel-good factor that animal charities like, and if it helps them raise funds for animal welfare, then perhaps there's no harm done. However, this degree of personification of animals has risks – most significantly, that the general public will unwittingly be persuaded that their pets think just like humans do. Surely the use of the word 'heroism' implies that the animal has consciously decided to take a brave course of action with the potential for injury, and rejected a cowardly but safe alternative. Yet none of the recent research into how dogs' (or cats') minds work has so much as suggested that they might be able to make such deliberate choices. They appear to live almost entirely in the present, reacting to events as they unfold, never stopping to reflect on which course of action might be the safest. Tara was probably just a feisty cat reacting instinctively to the snarling of the neighbour's dog. Obi the police dog was doing what he'd been trained to do (evidently, very well), and very sadly he was seriously injured as a consequence. If Obi had un-derstood that he had alternatives – either grab a rioter and have his skull staved in with a brick or bark ferociously from around the edge of his handler's riot shield – which do you think he'd have chosen?

Routinely – and often without any conscious thought – we speak of animals as if their brains worked the same way as ours. This habit stems from deep within our psyches. We speak this way equally about animals that are 'part of the family' and those we have never met; we also apply similar thinking to anything around us that appears to have some kind of life of its own – robots, computers, even the weather. This could reflect laziness on our part, compounded by a dearth of language specific to the description of animal minds – we call pets 'family' because we lack an alternative word.

The legend of Greyfriars Bobby shows just how prepared people are to believe that 'man's best friend' is almost human. The story goes that Bobby, a Skye terrier, refused to leave the Edinburgh grave of his former master – a policeman called John Gray – for no less than fourteen years (see nearby illustration). In 1873, the year after Bobby (supposedly) died, a statue and fountain were erected in his honour, both of which stand to this day. The inscription reads, 'A tribute to the affectionate fidelity of Greyfriars Bobby. In 1858 this faithful dog followed the remains of his master to Greyfriars churchyard and lingered near the spot until his death in 1872.' It now seems likely that 'Bobby' was in fact a stray belonging to the curator of the cem-etery, who fabricated the story in collusion with the owner of a local hostelry to attract and then profit from sentimental tourists – indeed this ploy was so successful that when the original 'Bobby' died after nine years, a similar dog was substituted to keep the scam going for a further five.[2]

Our projection of minds onto animals and inanimate objects alike is the expression of a psychological framework that we evolved to cope with uncertainty in the many forms it presents itself: what an animal is going to do next, when the next storm will arrive, whether the winter will ever end. Before science, it was our best way of deal-ing with everything unpredictable and uncontrollable. According to

Greyfriars Bobby.

one theory, it even gave rise to belief in the supernatural and hence triggered superstition and, subsequently, religious belief. Hence, while not unique to our dealings with pets, anthropomorphism is an essential component of how we relate to them.

We ROUTINELY AND almost without thinking speak about the world around us as if it had thoughts, feelings and intentions, even when a moment's reflection might easily persuade us that this is implausible. Broadly grouped under the umbrella (and tongue-twisting) rubric of 'anthropomorphisms', such self-evident errors go virtually unnoticed in everyday life. Most of us have at some time begged our cars to start on a cold, wet morning, even though a nagging voice in the back of our minds reminds that the car is blameless; rather, we should have taken the trouble to replace the battery before winter set in.

When an office computer slows to a crawl or crashes, we castigate it shamelessly for its errant ways, even in front of our co-workers.

Other times we may ascribe words or emotions to the 'minds' of machines simply due to the biases embedded in our language. Such so-called weak anthropomorphism occurs when we wish to offload responsibility for our actions onto some inanimate object, but we don't expect the object to talk back. It may make us feel better to shout at our computer when it's just 'lost' a file that we neglected to back up, but we don't for a minute think that we have hurt its feelings. Alternatively, we may simply be using everyday language to describe something efficiently, expecting other people to understand because we know that their imaginations work roughly the same way as ours do.

The 'strong' form of anthropomorphism occurs when we automatically ascribe to objects or animals minds of their own. This is also highly instinctive, triggered by more or less hardwired mechanisms in our brains. For example, anything that looks remotely like a face immediately grabs our attention, activating a specialized area in the visual part of our brain, the fusiform face area. But it takes more than a pair of eyes and a mouth to convince our brains that something is alive. That requires movement, particularly movement that appears purposeful. The classic 1944 experiments of Fritz Heider and Marianne Simmel perhaps awoke science to this phenomenon for the first time. They showed their subjects short animated films that depicted simple geometric shapes (triangles, circles, squares) interacting in apparently intentional ways (see nearby illustration). For example, one person described the sequence shown as follows: 'The large triangle chases the small triangle while the circle watches nervously from the door of the house', imputing both intention and emotion to simple geometric shapes that could not possibly beget either. Another subject demonstrated even greater anthropomorphism: 'The two men have a fight, and the girl starts to go into the room to get out of the way and hesitates and finally goes in. She apparently does not want to be with the first man.' On the surface, this description could represent an anthropomorphic use of language, not a serious attribution of human thoughts and feelings to objects in a film. Perhaps the way an object moves sometimes parallels the way people behave, allowing us to describe its actions using words that we can assume others will understand or that allow us to be more concise. Contrast the two examples given with that of the only subject in the study to use the more logical language of geometry to describe what she had seen: 'The circle enters the rectangle while the larger triangle is within. The two move about in circular motion and then the circle goes out of the opening and joins the smaller triangle which has been moving around outside the rectangle.' Anthropomorphic descriptions can be gratifyingly succinct as well as, somehow, more fun.

Subsequent experiments using a variety of similar movie clips have confirmed that anthropomorphism entails more than an economical use of language. Our brains contain two specialized 'modules' for

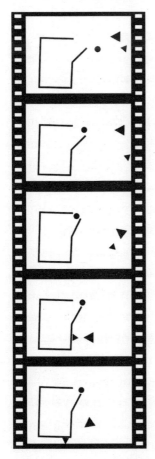

detecting and analysing any movement that might indicate intentional (that is, human) behaviour, even in interactions between simple shapes. Modules such as these – which may serve many other purposes, such as decoding speech and facial expressions – are not necessarily located in a single area the brain, just as the fuel system (tank, pump, injectors) and the sound system (radio, front loudspeakers, rear loudspeakers) are distributed throughout a car.

The first module, known as the mirror neuron system, primarily analyses actions performed by other humans by constructing simulations of those actions within our brains. More general in its scope – and hence more relevant to anthropomorphism – is the second module. Called the social network, this

Frames from the animated cine film made by Heider and Simmel.[3]

module compels us to describe the movements of geometric shapes as if they were people. The amygdala, which processes emotional and socially relevant information, forms one part of this – patients with damaged amygdalae describe the movie clips in exclusively mechanical language. However, brain scans of people watching the movies indicate involvement of at least six other parts of the brain; some analyse intentional motion, whereas others probe for the presence of other minds (mentalizing). We begin anthropomorphizing in situations where the entity we are addressing seems likely to have humanlike thoughts and feelings, but we do so most fully when we also believe that it is thinking about us.[4]

In some situations our anthropomorphizing is little more than a convenient habit. The computer I'm using to type this book, and at which I occasionally curse, stores all manner of information on my behalf, far more than I would wish my brain to hold, but I don't believe it holds opinions about me or feels upset (or relieved?) when I click 'Shut down' at the end of the day and it quietly mutters to itself for a few moments before turning itself off. By contrast, I, like most pet owners, have always talked to my dogs and cats as if they could understand me and interpreted their responses as if they have grasped something of what I said – perhaps not the literal meaning but at least the emotional content.

As computers become ever more sophisticated, imbued with artificial intelligence, we might predict that our weak anthropomorphism towards them might morph into a stronger form. Indeed we can see this happening in robotics, specifically in how people conceive of the robots with which they interact. As human populations become increasingly elderly, concerns are emerging as to who will tend to the needs of the infirm. Japanese engineers are developing 'care robots' as a potential solution. One issue in the design of such robots entails making them acceptable to their often uneasy clients. It turns out that mechanical-looking robots are acceptable, as are those that resemble animals – witness the popularity of the relatively simple PARO robot, modeled on a baby harp seal. People often reject as 'creepy' those that look somewhat like humans: this may reflect a tussle in the mind of the perceiver as to whether the robot can make judgements about them – is this man or machine? – indicating the point at which anthropomorphism stops and attempts at mind reading, normally reserved for members of our own species, begin. Thus our psychological mechanisms for dealing with animal behaviour are not simply watered-down versions of those we use when thinking about each other. We can clearly distinguish between situations that warrant one or the other – until we encounter an unconvincing android that is not quite either.[5]

SUPERSTITION – AND BELIEF in the supernatural more generally – is an extreme form of anthropomorphism. Nowadays we don't generally treat pets like gods, so the connection may not be immediately obvious, but pets are, to a certain extent, imaginary constructions. Understanding how our imaginations create completely intangible beings should illuminate how we think about those actual beings whose minds are merely somewhat inaccessible.

Philosophers first noted the connection between our imaginations and the supernatural thousands of years ago; the Greek theologian Xenophanes coined the term 'anthropomorphism' in the sixth century BCE. Even today, in an age dominated by technology and science, around one-third of Americans admit to a firm belief in ghosts, and less than half are sure that they don't exist. Many people not only believe in haunted houses but actually seek them out when on holiday. According to the front desk manager at the Concord (Massachusetts) Colonial Inn, 'Room 24 is the big attraction – we call it our "room with a view-ing". It's haunted, according to local lore – and to the many guests who have witnessed strange happenings while sleeping in it. It's always either/or. People either ask for the room specifically or ask to be put far, far away from it.' In the United Kingdom, over 60 per cent of women and over 40 per cent of men believe that some people (not necessarily themselves) have experienced ghosts. About one in five Brits and 50 per cent of Americans believe that a person can become possessed by the Devil or another evil spirit.[6]

Although we in the West live in increasingly sceptical times, across the globe those who profess a religion greatly outnumber those who don't. Even in the United Kingdom, more than half the population believes in a deity or at least some kind of supernatural power; in the United States, that number is nearer to nine out of ten. Few politicians dare to profess that they are unbelievers (although in 2010 two leaders of major UK parties, Nick Clegg and Ed Miliband, did admit to being atheists; both are now backbench members of Parliament).[7]

Scientists increasingly believe that our acceptance of the supernatural has evolutionary origins. According to one theory, such beliefs functioned to bring social and moral cohesion to extended families. To leave as many descendants as possible, grandparents should do everything possible to prevent rivalry among their offspring and promote their survival – if necessary, at the expense of neighbouring (unrelated) families. Throughout much of human evolutionary history, we lived in groups of between 50 and 150 individuals, mostly comprising a single biological family with a sprinkling of in-laws from nearby tribes. The assertion of a shared behavioural code was essential to maintain order, especially as these groups grew in size. According to this theory, ancestor worship – common to this day in many small-scale societies – would have emerged organically in this context. It would have been but a short step from 'Do this because Grandad says so' to 'This is what your grandfather would have told you to do when he was alive.' Shamans, who served as repositories of cultural knowledge and moral strictures, and solved day-to-day social problems by interpreting ancestral codes of behaviour, could then mediate connections between living members of the community and long-dead forebears. Shamanism also persists to this day in small-scale societies, such as the Hopi of the southwestern United States and the Australian Aboriginals, and as an element of the success of breakaway sects such as those of the Maharishi Mahesh Yogi (briefly embraced by several Beatles) and the Reverend Jim Jones, members of whose 'People's Temple' committed mass murder/suicide in the jungles of Guyana in 1978.[8]

The largely cultural routes by which ancestor worship and shamanism evolved into organized religions have less to tell us about our
relationships with animals. Nevertheless, the absence of any totally
atheistic societies has prompted psychologists, biologists, anthropologists and theologians to postulate that religious belief is intrinsic to
human nature.

It is one thing to point to the ubiquity of religion; quite another
to identify the psychological mechanisms that underlie it. There can be
no 'gene for religion': genes make proteins, or influence how other
genes make proteins, and cannot on their own create a particular
way of thinking. Logically, any mechanisms that predisposed belief
(in general) must have their basis in genetic influences on our brains'
construction. Nor can there be genes for any particular religion: the
faith a person adheres usually depends on the culture into which he
or she happened to be born.

Acceptance of the supernatural begins with the psychological
mechanisms that attribute random events or otherwise inexplicable happenings to an otherworldly being that not only caused it but
also had us in mind when it did so. Evidently we feel compelled to
blame someone or something when misfortune strikes. A credible
supernatural agent not only serves as the cause but also provides an
explanation – for example, the misfortune is a punishment for some
past misdeed; supplication to that agent can reduce anxiety that the
event might recur. Belief in an afterlife of the mind also appears to
have a biological basis, since it appears early on in childhood. In one
study, preschool children shown a puppet show in which a crocodile
ate a mouse agreed that the mouse would no longer need to eat but
believed that it might still get hungry – that its mind lived on even
though its fuzzy felt body had disappeared.[9]

THREE PSYCHOLOGICAL MECHANISMS seem to underlie our propensity to believe in supernatural agents – spirits that not only observe
what we do but actively steer our lives. The first is our tendency to
imagine intention in everything that happens to us, especially the
unpleasant. Nowadays we protect ourselves from harm by all manner

of technologies (alarm systems) and social contracts (professional police officers). However, for much of the millions of years of the human brain's evolution, our hominid ancestors faced all kinds of dangers, including predators that could run considerably faster on four legs than they could on two, as well as other members of their own species intent on making off with not only their possessions but also their scalps. Because detecting danger as early as possible was important to survival, our brains evolved specialized mechanisms for doing so, and because the cost of failure was so high, those mechanisms became hypersensitive. How many times has the smoke alarm in your home gone off while you were making toast or grilling a steak? Plenty, if your household is anything like mine, but I make sure that the alarm always has a fresh battery because I don't want it to fail the one time that the house really does catch on fire in the middle of the night. Just like a smoke alarm, the agency-detection system in our brains errs on the side of false positives, such that it's technically referred to as the hyperactive agency-detection device (HADD).[10]

We're not simply inclined to detect malice wherever it might possibly occur; we are also programmed to try to out-think whatever agent might pose the danger. This second mechanism derives from the reason our big brains evolved in the first place: to predict what other people were thinking before they could harm us. This theory of mind, or mentalizing system, enables us to consider other people's thoughts, not simply react to their behaviour as most animals do. It's such an intrinsic part of human nature that we do it all the time, effortlessly. As soon as the HADD notices that the action of some kind of mind might possibly have caused some event, the mentalizing system swings into action and generates hypotheses about what that mind might be thinking. When we wake in the middle of the night, such chains of logic as 'I think I heard a noise downstairs, it could be a burglar, if I go down and confront him he'll probably attack me' flash through our minds before we're even halfway alert.[11]

Put together, the HADD and mentalizing system provide a basis for superstition and belief in the supernatural. A sudden storm drives a primitive boat full of valuable cargo onto shore, and the mariners barely escape with their lives. They must then decide when to venture out again but have no logical basis for deciding how long to wait, so resolve their indecision by spontaneously inventing a supernatural agency that controls the weather. They can then attempt to manipulate the agency by sending prayers, offering gifts, or making sacrifices.

Because we find it difficult to imagine otherwise, we default to giving our supernatural agencies humanlike thoughts and intentions. In Charles Darwin's words, when developing religion humans 'would naturally attribute to spirits the same passions, the same love of vengeance, or simplest form of justice, and the same affections which they themselves feel'. As religions mature, they evolve more codified and abstract representations of their deities, but many individual believers still talk about their deity as if it were little more transcendental than we are. This from one study: 'God was listening to two birds singing in a tall tree next to an airport. When a large jet landed, God listened to it because he could no longer hear the birds. Then he listened to the birds again.' At the personal level, anthropomorphism of deities sometimes differs little from anthropomorphism of pets.[12]

A third factor in our tendency towards supernaturalism may be our instinct to look for function and connection in everything we observe: 'a time to every purpose under the heaven'. This is most evident in children: most four-year-olds identify clouds as 'things for raining' and only learn gradually that more scientific explanations exist. For example, when asked to choose between two answers to the question 'Why did the first ever thunderstorm occur?' about one-third of British six- and seven-year-olds chose the answer 'The first ever thunderstorm occurred to give the earth water so everything would grow' rather than 'The first ever thunderstorm occurred because cold and warm air all rubbed together in the clouds'. In answer to 'Why did the first ever bird exist?' more than half preferred 'The first ever bird existed to eat worms and insects so there wouldn't be too many of them' over 'The first ever bird existed because an animal that lived on the ground began to develop wings and fly', showing that that they more easily ascribed purpose to animals than to the weather.[13] Learning seems to overlay this way of thinking but not

suppress it entirely: even some scientists, pressured to give a snap answer, endorse such statements as 'The Earth has an ozone layer in
order to protect it from UV light.'

Combining all three biases, it's but a short step from divining intention in small events, such as imagining an intruder to have caused
a noise in the night, to blaming supernatural agents for inexplicable
catastrophes (as Ray Nagin, mayor of New Orleans, supposedly did
in the aftermath of Hurricane Katrina: 'Surely God is mad at America. He sent us hurricane after hurricane after hurricane.') Precisely
how these three mechanisms interact remains unclear but may eventually be revealed by studies of the brain. One functional magnetic
resonance imaging (fMRI) study has already suggested that the in-
struction to think about God activates all the parts of the brain associated with action detection (the HADD), person perception and
mentalizing.[14]

Advertisers understand well the power of these biases to alter
human behaviour. Car manufacturers cunningly design car grilles to
look like faces: they use anthropomorphism to break down our rational self-control. We may know that high-calorie snacks, however
tasty, are bad for us, but giving them human characteristics can apparently cause some people's resistance to crumble. Who could resist
the charms of Mr Red M&M, who appears in the brand's advertising:
'Age – He says thirty-something, but in fall 2010, an M&M commercial said 46: Weight – Perfect for his shell size: Turn-ons – When
people blindly follow his wise advice: Turn-offs – When people fail
to recognize his obvious leadership abilities.'[15]

WHILE WE HAVE some way to go in working out the details of the
precise biological underpinnings of religion, they do provide some
useful clues as to how people might instinctively think about animals in general and pets in particular. Many animals' movements as
they try to escape from us or the postures they assume when thinking
about pouncing on us, even the rustling sounds they make as they
move around, should trigger the HADD. Unlike ghosts, animals react to our presence in a sensible but not entirely predictable way,

encouraging us to think of them as capable of mentalizing. Notably, even in those studies primarily focused on the psychology of religion, the statements involving animals triggered anthropomorphic responses from the greatest number of subjects.

Obviously we don't automatically presume that everything that moves is thinking about us. However, we do tend to fall back more on such explanations under some circumstances than others. For example, we tend to be more logical about positive or pleasant occurrences than about negative or unpleasant events. So we're much more likely to curse at our computer when it crashes than to shower it with praise when it does what we want it to; conversely, people who regularly have computer trouble are much more likely to attribute feelings and intentions to it than those whose computers work well most of the time. We also anthropomorphize more when we're stressed than when we're calm, probably because stress inhibits the more rational, scientific explanations that would otherwise suppress the more primitive attributes of evil intent.

Our ability to suppress our instinctive anthropomorphism grows as our brains mature. As well as ascribing to everything a purpose ('tigers are for looking at in the zoo'), young children, whose experience of the world has been almost entirely social, treat all living things as if they had minds just like their own. (Indeed, even when we are adult and can put ourselves in other people's shoes, we must sometimes still suppress our instinctive egocentric interpretations.)[16]

As children learn more about how the world works, their naive interpretations gradually give way to the 'correct' ones, shaped by the science-based knowledge they acquire or everyday experience. Thus rural children who have more often encountered the realities of life and death are less sentimental about the animals that form part of their everyday experience than are urban children. Indigenous Mayan children from the Yucatan absorb their parents' 'folk biology' from an early age and from five onward begin to describe animal behaviour in its own, non-human terms – especially the boys, once they start accompanying their fathers on hunting trips into the forest. As adults, we consider meteorological explanations for storms supplied to us by professional weather forecasters more reliable than our forefathers' attribution of them to an unseen being in the sky. But as soon as we leave the comfort zone provided by our limited store of technical knowledge, anthropomorphism stands at the ready to give us the explanations that our minds crave.[17]

People use anthropomorphism not only when outside their comfort zones but also when attempting to increase their control over unpredictable or unexpected occurrences – from major events such as hurricanes and deaths to more minor irritations such as machines that don't perform as expected. One study asked participants to place metal ball bearings on a grid of small holes in a board. Unbeknownst to them, magnetic fields, occasionally switched on underneath the board, caused the ball bearings to jump from one hole to the next. In subjects' running commentaries on what was happening, the unexpected movements prompted remarks such as 'That one did not want to stay', 'Oh, look. Those two kissed', and 'They are kind of fighting'. In another study, subjects memorized a list of gadgets, some described as unpredictable and others as predictable; they then entered an fMRI scanner and were shown pictures of the gadgets. The areas of the brain associated with mentalizing (the medial prefrontal and anterior cingulate cortices) were among those specifically activated when they saw the unpredictable gadgets.[18]

Feelings of loneliness or depression are particularly powerful in enhancing anthropomorphism. As members of a highly social species, most of us have a need to belong (one exception being people on the autistic spectrum, who notably do not often anthropomorphize). In one study those who reported feeling chronically lonely tended to describe their family pets as thoughtful, considerate and sympathetic – all qualities helpful in bridging the species barrier. However, they were no more likely to project other, sometimes negative human qualities onto their pets – for example, a capacity for embarrassment, creativity, deviousness or jealousy. A related study showed participants excerpts from the movie *Cast Away* to induce vicarious feelings of loneliness, and their ratings of their family pets then changed in favor of precisely those qualities that might provide emotional support (viewing scary clips from *Silence of the Lambs* produced no such change, so the effect seemed quite specific to loneliness). Thus in both studies the subjects seemed to pick out specifically those imaginary humanlike characteristics that matched their needs at the time, but they did not portray the animals as more humanlike overall – displaying a selective version of anthropomorphism. Similar shifts occur in people's religious behaviour. Isolation may make prayer more satisfying: in a 2005 poll of Christians in the United States, 40 per cent indicated that they felt closest to God when they were alone, and only 2 per cent said the same about being in church. Among dog owners, anthropomorphism is most intense in those who feel depressed or socially anxious.[19]

While attributing thoughts and feelings to objects and animals seems intrinsic to human nature, not everyone is equally susceptible. Some people are more prone than others to resort to anthropomorphism across a range of situations. For example, people who believe that their dog understands them are also likely to think that some kind of silicon-based spite motivated their misbehaving computer. Brain scans link this general tendency to enlargement of one very specific part of the brain (the left temporoparietal junction), an area involved in mentalizing. It's unclear whether this increased size results from constant anthropomorphizing or is congenital and enables those with it to anthropomorphize more freely than others. In either case, the effects are wider ranging than simple differences in the language people use to describe their pets.[20]

People prone to anthropomorphism conceive of living things quite differently from those who are more hardheaded. They tend to invoke a moral dimension when confronting hypothetical decisions, such as whether it is right to destroy a high-powered computer. They show greater concern for conservation of the natural world, possibly because animals dominate their conceptions of the 'natural'; anthropomorphizing nature itself – even rather crudely, for instance by depicting the Earth as possessing a face – can strengthen their environmental concerns even further. They are more likely to believe, in interacting with machines or animals, that those entities are evaluating and judging them and, as a result, tend to behave more responsibly.[21] They are also much more willing to credit animals with experiencing complex emotions, such as admiration, resentment, shame, remorse, embarrassment, guilt, hope, nostalgia, humiliation and optimism – all of which rely on a greater degree of consciousness than science can support in any animal with the possible exception of some primates. Most biologists now agree that mammals – indeed all vertebrate animals with the partial exception of fish – can experience the full range of basic emotions, or 'gut feelings', including fear, anxiety, surprise, suffering, anger, affection, pleasure, and the feelings associated with sexual attraction and pain; those that require a degree of reflection, however, are almost certainly beyond the capabilities of all animals, with the possible exception of apes. Less anthropomorphizing people tend not to ascribe higher emotions to their pets, but those who do may put their pets in jeopardy by over-anthropomorphizing them. Owners who presume an imaginary capacity for complex emotions in their pets and other animals are more likely to misunderstand them and, as a result, to unwittingly maltreat them (see box on p. 178).

Importantly, however, anthropomorphizing animals is not the same thing as automatically projecting one's own emotional tendencies onto them. In one study, individuals who tended to feel anxious most of the time were no more or less likely than other people to interpret a dog's behaviour as motivated by anxiety. Those who tended to feel guilty more than average were not only more inclined to impute guilt to dogs but also portrayed the dog as more anxious. Those who were feeling lonely themselves were *less* likely than others to believe that their dog was feeling lonely. Thus while anthropomorphism offers one way for owners to discern emotions in their pets, a certain amount of objectivity prevents them from treating their animals as simply extensions of themselves.[22]

We may imagine that the sentimental perception of pets is a modern affectation, but the presumption of humanlike qualities in animals is far from a recent phenomenon: anthropomorphism of animals was evidently already rampant when Xenophanes coined the term. Indeed, he used the example of animals to satirize the gods worshipped by his sixth century BCE contemporaries as projections of their own personas: 'But if cattle and horses and lions had hands, or could paint with their hands and create works such as men do, horses like horses and cattle like cattle also would depict the gods' shapes and make their bodies of such a sort as the form they themselves have.' The tradition of zoomorphism (the depiction of a human with some of the characteristics of a non-human animal) continues to this day in all forms of art and has reached new levels in animated movies – although few people take the blurring of non-human and human animal natures so far as the self-styled 'furries', who attempt to assume the identities of various alternate species, including dogs and cats.[23]

When taken to extremes, anthropomorphism has occasionally had dire consequences for animals. In continental Europe, the justice system in the Middle Ages held some animals accountable for the 'crimes' they had committed; occasionally, indeed, formal trials took place. To quote an example from France:

> On the 5th of September, 1379, as two herds of swine . . . were feeding together . . . three sows, excited and enraged by the squealing of one of the porklings, rushed upon Perrinot Muet, the son of the swine-keeper, and before his father could come to his rescue, threw him to the ground and so severely injured him that he died soon afterwards. The three sows, after due process of law, were condemned to death; and as both the herds had hastened to the scene of the murder and by their cries and aggressive actions showed that they approved of the assault, and were ready and even eager to become *participles criminis*, they were arrested as accomplices and sentenced by the court to suffer the same penalty.[24]

Pragmatism triumphed in the end – the Duke of Burgundy pardoned the 'accomplices'. But the episode illustrates the extent to which people of the era deemed the behaviour of animals humanlike not so much in their attribution of murder to the so-called perpetrators but in their presumption that the pigs had acted as a wilful mob, cognizant of their actions and responsible for goading the 'murderers' into action – something our modern understanding of animal behaviour would not countenance today. (Yet even nowadays packs of marauding dogs sometimes get portrayed in much the same way.)

Anthropomorphism also allows us to use animals as metaphors. Throughout history, we have assigned specific human traits to different kinds of animals. Hyenas are treacherous, crocodiles cunning, ants industrious, turtle-doves peace-loving; dogs are loyal (or reproachful), while horses are proud. Some animals assume different characters according to context: cats, for example, may be magical, aloof, lascivious, devilish or free-spirited. Such anthropomorphisms provide a near-inexhaustible repository from which writers and filmmakers can draw when wishing to invoke a particular human attribute, confident that their audience will not only accept without question that the animal genuinely possesses that attribute, but know precisely which is being alluding to.[25]

Thus our anthropomorphism of animals is a result of how our brains are constructed. It lies at the heart of the superstition and shamanism that were the prehistoric equivalents of religion. We have every reason to suppose that our ancestors were imputing human thoughts to the beasts around them since our species first evolved. Although perhaps exaggerated by mawkish indulgence, anthropomorphism is not therefore the product of sentiment. Yet it has made pet keeping possible: had we not been inclined to imbue our pets' minds and actions with humanlike characteristics, we would have had no other way of understanding them, since scientific understanding of cats' and dogs' minds has only come about recently and is still not universally appreciated. It is difficult to see how, without anthropomorphism, a bond between humans and their animal companions could have arisen.

SOME PEOPLE ARE more likely to treat their pets as 'little people', but not all kinds of animal are equally easy to endow with human characteristics. Surprisingly, the precise features of animals that make them more or less susceptible to anthropomorphizing have not received a great deal of study. Historically, most biologists have identified anthropomorphic descriptions of animals' behaviour as errors, the product of an unfortunate glitch in human cognition that stands in the way of 'objective' analysis (although those who study primates sometimes argue the opposite). Perhaps as a consequence, scientists have neglected to explore precisely what tempts us to anthropomorphize in the first place.[26]

There have been many studies, however, of which animals we're likely to anthropomorphize – in large part, because the subject is very relevant to advertisers. Animals appear in many advertisements, sometimes obviously anthropomorphized and sometimes not, although their inclusion is inevitably symbolic (except in promoting pet products). In one 1993 survey of print ads from the United States, four types of animal – dogs, cats, horses and birds – predominated. Horses usually appeared in their natural state and were used mainly to advertise tobacco and alcohol, possibly to emphasize strength, potency, manliness and vigour. Cats were less common outside the pet products sector and, when not of the domesticated variety, were heavily anthropomorphized to make them appear less threatening and hence more likeable: a real tiger would be unwelcome at most breakfast tables but becomes attractive when transformed into a cartoon character, Tony the Tiger, enjoying a bowl of cereal (despite the fact that real tigers have difficulty digesting both grains and milk). Despite its popularity from the 1950s to the present day, this depiction of the tiger did not monopolize children's perceptions of this species. In the classic 1968 children's book *The Tiger Who Came to Tea*, the eponymous big cat is so hungry that he eats all the food in the house, washed down with 'all the milk, and all the orange juice, and all Daddy's beer, and all the water in the tap', although he is otherwise friendly and walks upright rather than on all fours. Products aimed at adults might be associated with animals in their natural wild state – for example, the cougar and jaguar create an untamed, powerful image for the cars named after them – or heavily anthropomorphized, as in the hipster cartoon camel 'Old Joe', used to advertise Camel cigarettes (whose purchasers were presumably unaware of the aroma of a real camel). The Old Joe character was discontinued in 1997 after the discovery that he was as appealing to children as he was to adults.[27]

When it comes to effective advertising, appearance is important: more human-looking species get anthropomorphized more. Overall, mammals tend to beat birds, which in turn beat most reptiles; many people think of fish, amphibians and crustaceans as little more than automata, although even insects can be anthropomorphized in very specific ways: honeybees are sometimes portrayed as possessing a collective intelligence that approaches that of humans. However, insects generally elicit little concern when it comes to conservation, where a perceived similarity to humans is an important motivator: when researchers asked the participants of one study to choose one endangered animal to save from a list of six threatened species, the gorilla beat the rhino, which beat the only bird in the list (a crane), which in turn beat a lizard and a catfish. The Tooth Cave ground beetle (found only in a handful of caves near Austin, Texas, and the nearest to extinction) was almost universally rejected.[28]

Faces are important, and within faces, eyes. It's no coincidence that the emblem of the World Wildlife Fund (now simply WWF) is a panda, an animal whose face resembles a human in clown makeup, which has the effect of making its eyes look huge. When looking at their dogs' faces, owners tend to concentrate on their eyes – the telltale of the human face – rather than their ears, which better indicate the dog's intentions. Eyes seem to be the most important feature in the supposed similarity that owners share with their dogs – if true, this implies that when choosing a dog, many owners subconsciously pick those breeds that look facially most like their relatives (or themselves). Making a dog's iris more distinct – subtly more humanlike – on a photograph can enhance the animal's appeal. A mouth that we can imagine as smiling is also attractive – even though the 'grin' on a dog's face expresses not so much happiness as a desire for acceptance. Expressions that look especially humanlike often strike a chord. In one study, dogs that raised one eyebrow (as if quizzical) at potential adopters were rehomed faster than those that didn't, even than those that wagged their tails and came to the front of their pens, factors already known to make a dog more appealing.[29]

Movement also seems to be a key trigger. Although biologically speaking both animals and plants are living entities, we usually restrict the lay definition of 'alive' to animals, at least in the West (other cultures may differ: for example, the Itza Maya of Guatemala consider various trees, as well as the sun and the 'spirits of the forest', equivalent biological entities to a peccary or a turkey). Plants occasionally receive recognition, as in 1983, when the newly discovered ability of some trees to '(l)eavesdrop' on one another (detect that their neighbours are under attack from insects) prompted an outpouring of anthropomorphic puns, such as 'Scientists Turn New Leaf, Find Trees Can Talk' (Los Angeles Times) and 'talking trees whose bark is worse than their blight' (New York Times).[30]

The speed at which an animal moves is especially important – too fast, and people are less likely to project humanlike qualities onto its behaviour. Animals that 'scurry' are 'scurrilous'; if they're too slow, like a tortoise, perhaps they seem too plantlike. People tend to describe as intelligent and intentional those animals that move at roughly the same speed as humans (see nearby illustration). But within those limits, it seems, the relative, not absolute, speed is important. In an especially neat experiment, a team of psychologists asked people to watch animated movies of a blob devouring vehicles and street furniture while humans performed normal activities nearby. If the animation of the humans was speeded up – or slowed down – relative to that of the blob, the blob's 'intelligence' plummeted.[31]

The domestic dog and cat stand out clearly from the other species in the illustration depicting the attribution of minds to non-human animals. Dogs typically get rated as possessing similar mental abilities to gorillas and other non-human primates, superior to those of their ancestor, the wolf (despite the evidence that as wolves became domesticated into dogs, their brains shrank). Cats, while not rated as highly as dogs for intelligence, significantly beat their big-cat cousins, including the lion and the cheetah (although a biologist

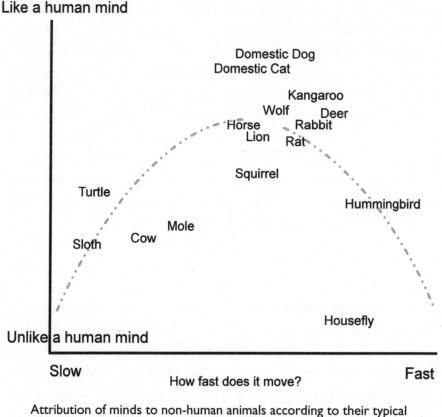

Like a human mind

Domestic Dog
Domestic Cat

Kangaroo
Wolf   Deer
Horse   Rabbit
Lion   Rat

Squirrel

Turtle   Hummingbird

Mole
Sloth   Cow

Housefly

Unlike a human mind

Slow   Fast
              How fast does it move?

Attribution of minds to non-human animals according to their typical
speed of movement. Note the anomalous position of the domestic dog
and cat, both rated as more intelligent than their wild counterparts.

would say that the brains of all three are remarkably similar). Familiarity alone may not be the deciding factor, since neither the rabbit nor the horse seems to gain any advantage over their wild counterparts, so the crucial attribute may be the dog's and cat's status as 'part of the family'.

We also anthropomorphize different kinds of pets differently. People generally do not ascribe complex emotions, those that require self-consciousness, to non-human animals, or do so hesitantly. Thus while anthropomorphism itself has biological foundations, cultural factors may strongly influence its precise expression. One survey of pet owners conducted in the United Kingdom revealed a general consensus that mammals and birds share a common repertoire of basic 'gut feelings': sadness, anger, affection and fear, to name a few. There was also general agreement that rodents and birds are unlikely to experience any of the more complex emotions, including embarrassment, shame, guilt and empathy. However, around half of the people surveyed – all of whom were long-term pet owners, mostly of cats and dogs – thought that those species could experience guilt, empathy, pride and jealousy. Moreover, they associated three of these emotions with particular species (see nearby illustration): guilt with dogs, pride with horses, and jealousy with both dogs and horses (but not cats). Researchers have found a similar pattern for dogs among Hungarian dog owners.[32]

A STRANGE PARADOX is at work with pet ownership. People are likely to know a great deal more about the species they own than any other. This knowledge should make them less, not more anthropomorphic towards them – just as understanding meteorology makes us less susceptible to believing that malevolent spirits cause storms and droughts. We have come to rely on a rational explanation – physical

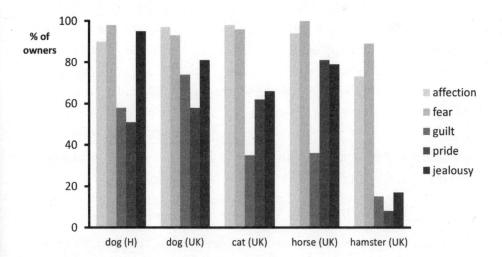

Belief among animal owners in the United Kingdom and Hungary that dogs, cats, horses and hamsters are capable of experiencing selected basic emotions (affection, fear) and complex emotions (guilt, pride and jealousy). The first set of columns depicts similar beliefs among dog owners in Hungary (H).[33]

factors such as temperature and the rotation of the Earth drive weather systems – even though in parts of the world with unstable weather, the forecasts are far from reliable.

By contrast, pet owners are much more certain than non-owners that animals (in general, not just their own) have emotional lives. It's unclear whether this is the consequence of living with and growing fond of an animal or is a factor that drives people to obtain their first pet: if (hypothetically) you thought that animals didn't have much in the way of feelings, why would you bother to spend all that time and effort looking after one? Dog and cat owners consistently overestimate their pets' other human qualities and mental capacities. Somehow, having a personal relationship with an individual animal seems to involve imbuing it with characteristics that science would restrict to our own species.[34]

A heightened sense of anthropomorphism may not be essential to a satisfying relationship with a pet, although some degree of it is undoubtedly important. In one study of cat owners, I examined the extent to which they anthropomorphized their pets' behaviour and compared this to the amount of emotional satisfaction they obtained from the relationship. Somewhat surprisingly, to me anyway, I couldn't find much of a link: only those people who doubted that their cats had any emotional life at all – presumably, those who subscribed to the notion that cats are calculating, cold, heartless beasts – didn't take much comfort in their pets. Whether the same is true of dog owners remains to be seen.[35]

ANTHROPOMORPHISM EMERGES FROM an inbuilt feature of the human mind that enables us to deal with the world around us and to build and develop relationships with non-human animals. Our initial reaction to the unpredictable, a tendency to attribute agency seemingly hardwired into our brains, gave rise to a human-centred way of coping with reality to which even the most hardheaded of us is not entirely immune. In keeping with these rapid and unnuanced judgements, we then fall back on simple rules of thumb to judge whether an entity is likely to be thinking about us, be it a robot, a wraith or a rat. Moreover, we tend to default to a simple scale that runs from mechanical, through infantile ('Dogs have the mind of a four-year-old child'), to almost human.

We may never know whether our distant ancestors began by anthropomorphizing animals and then moved to inanimate objects or the other way around – or indeed precisely when the human mind first became capable of anthropomorphism at all. The evidence from archaeology (discussed further in Chapter 9) suggests that this might have happened any time between 60,000 and 30,000 years ago. We can be reasonably sure that anthropomorphism of animal minds enabled our ancestors not only to hunt more effectively but also to bring individual animals into their lives. Their newfound ability to guess what animals might be thinking would have enabled hunter-gatherers to capture baby animals and then raise them in their villages, the precursor to modern pet keeping.

Anthropomorphism still exerts a powerful hold on our ways of thinking about animals, despite the emergence of a more objective, science-based alternative. True, scientists have only recently begun to provide an understanding of the extent to which animals' minds differ qualitatively from our own, fitting the circumstances in which each species evolved, but their more rational explanations have not been widely adopted. Even more to the point, over the past two decades some of the greatest advances in our knowledge of animal minds have come from dogs; yet the extent to which dog owners treat their pets like little humans remains undiminished and may actually be increasing. In fact, of all animals we most anthropomorphize dogs and cats – the animals that live unconstrained within our homes. We seem driven to overestimate their intelligence, and many owners find it credible that their dogs might have a 'sixth sense' that enables them to predict the future (when their owners are going to come home). Thus, paradoxically, the more we get to know them, the more humanlike we want our pets to be.[36]

We do not, however, feel the need to care for everything we anthropomorphize. We prefer to confine our ghosts to fiction. Those who profess affection for their shiny laptops feel little compunction in regularly trading them in for even shinier models. Only our pets

become 'family'. Logically, there must be a reason why we still con-
ceptualize as 'little people' – perhaps more so than ever before – the
very animals about which we now know so much, both from everyday experience and from knowledge acquired from various authorities. Anthropomorphism, it turns out, has a powerful ally in our
emotional reactions: we also find our animal companions cute.

Anthropomorphism – a Dog's Perspective

If owners did not believe their pets were sentient beings, many of
them would probably never have been born. Only our increasing
habit of personalizing dogs and cats can account for their current
popularity. Yet, from the animal's point of view, anthropomorphism
has its downsides.

I'm often asked whether dogs (especially) feel degraded when
dressed up in comical costumes or carried around like human infants.
Since no evidence suggests that dogs can feel embarrassment – however much we may read that emotion into their behaviour – I always
answer with a cautious no. If the pup is comfortable, it's probably not
suffering.

Our misunderstanding of our animal companions runs much
deeper. The biggest mismatches between our conceptions and their
well-being stem from our failure to understand their emotional lives
and how they think. For instance, most dog owners believe that their
pets feel guilt, and many will punish their animals for whatever crime
they imagine has generated this emotion (e.g., scratching the furniture, peeing on the rug). However, Professor Alexandra Horowitz
blew that myth apart in a game-changing experiment demonstrating conclusively that a dog's 'guilty look' in fact represents an exqui-
site reading on the dog's part of the owner's body language at the
moment of discovering the misdemeanour.

The mistaken idea that dogs can feel guilt stems from a fundamental misconception about how their brains work. Unlike us, dogs
live almost entirely in the present, unable to plan ahead more than
a few minutes. They are not given to excessive reverie. Connecting

something they did several hours ago with their owner's behaviour now exceeds their capabilities. Under these circumstances, when punished, they don't understand why and hence make the wrong connections. This in turn renders them less confident in their relationship with their owner because they can't work out why they alternately get punished or welcomed with affection when he or she returns home.

Another common misconception is that dogs seek domination within the household. While children may jostle for status in the family, dogs certainly do not – despite what certain 'experts' say. With this flawed idea in mind, some owners punish their pets to 'keep them in line'. But this makes no sense from the dog's point of view. Their canine minds don't work like ours do: our brains evolved to understand and profit from social situations; theirs, to understand their environment (especially through smell, a sense we barely possess). Indeed, science has yet to demonstrate that dogs even appreciate that we are capable of thinking about them. Whatever message a dog receives every time it feels a sharp tug on its check chain, it won't be that it should have more respect for its status in the family.[37]

CHAPTER 7

How Cute Is That?!

FOR THE ANIMALS among us, cuteness counts. Cute dogs stand the best chance of getting adopted, while their not-so-cute kennel mates get put to sleep. Most would-be adopters spend less than a minute evaluating each dog, and most use its appearance as their main criterion. Puppies and younger dogs generally find homes more quickly than their less cute peers. People assume cute dogs will be more loving.

Cuteness is a powerful biological force that ensures that mammals will care for their offspring. We find our own infants and toddlers cute, and female cats and dogs likely feel the same about their own helpless offspring. We are unusual, however, in responding so powerfully to cute looks in other species, most notably in our pets but also in wild animals whose appearance triggers the same attraction (sometimes erroneously). The less inhibited of us spontaneously break into baby talk when we catch sight of a puppy or kitten. The supermodel cradling her toy dog bears obvious similarities to the stay-at-home mum with her new baby. Cuteness knows no species boundaries.

Albeit influenced by culture and experience, our perceptions of cuteness affect all our interactions with animals, first impressions especially. If anthropomorphism leads us to guess what our pets are thinking, their cute features draw us in to bond with them emotionally. Cuteness on its own, however, probably isn't enough to sustain the lifelong relationship that follows – a wheezy, smelly, shuffling old dog is not as cute as a puppy – so it cannot be the be-all and end-all, but it often plays a big part in a prospective owner's choice of pet.

Primarily, our responses to cuteness must have evolved to benefit our own species, to ensure the survival of our own offspring. Recognizing that mothers who fail to bond with their own offspring may somehow be oblivious to their cute features, science has paid much more attention to our responses to own infants' looks than to those of baby animals – although, somewhat presumptuously, in their experiments they have sometimes used pictures of baby animals as proxies for human babies. Thus before examining how we respond to cute animals, we need to examine our responses to human infants, which must have played an important role in our own survival.

NOBEL PRIZE–WINNING BIOLOGIST Konrad Lorenz first pointed out that we likely evolved our reactions to cuteness. He proposed that the way infants look instinctively drives mothers (and, in our own species, fathers) to care for them. During the evolution of the human species, mothers who failed to respond to their babies would have left fewer surviving offspring than those who did, so the response became genetically fixed (and it's still important today: excessive intake of narcotics or alcohol during and after pregnancy are powerful blockers of the cute response in women, more serious even than postpartum depression). The cuteness of babies affects how keen their mothers are to take care of them: in more than one study, mothers of less cute babies were more attentive to visitors to the maternity ward than they were to their newborns, whom they looked after in a mechanical rather than affectionate way. Cute children tend to be rated as friendlier, healthier and more trustworthy than their less cute counterparts. For us humans, as for our pets, cuteness matters.[1]

What exactly is cuteness? Lorenz thought that the key features, which he named Kindchenschema, or 'baby pattern', were likely to include

- A large head in proportion to the size of the body
- A relatively large forehead

- Large eyes, which, because of the large forehead, are slightly more than halfway down the head
- A small nose and mouth
- Round, protruding cheeks
- A rounded body
- Short, chubby limbs
- Soft, elastic skin

These are all static features, but the sounds that happy babies make – such as babbling and cooing – and their uncoordinated movements must additionally contribute to their cuteness. Those factors have received less study, however, because their impact is harder to assess: researchers can manipulate the first five factors in photographs of faces, making it much easier to compare reactions (see illustration below). That said, the relative importance of each feature on Lorenz's list has received little attention, apart from one early study (using cartoon faces) suggesting that the size and position of the eyes are most crucial.[2]

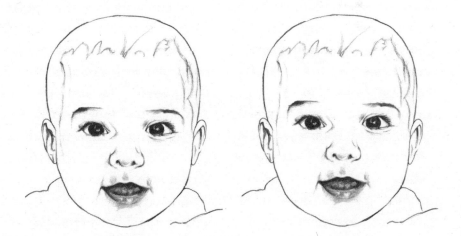

Seemingly trivial manipulations of images of babies' faces can reduce their perceived cuteness: the picture on the right is rated as marginally less cute.

Cuteness has profound effects on our behaviour. Faces with cute features grab our attention faster, make us smile involuntarily, prompt us to approach and utter exaggerated greetings in baby talk, and make us feel protective. While these reactions tend not to vary, their intensity does. Men are generally less outwardly responsive than women and less perceptive than women in judging fine differences in apparent cuteness in infant faces (an outcome seemingly specific to cuteness – for example men and women are equally adept when it comes to judging the emotion conveyed by unfamiliar infants' expressions). Hormones play a role: women of child-bearing age are more responsive to cuteness than prepubescent girls or postmenopausal women, as are women using oral contraceptives containing the hormone estrogen. Oxytocin, the hormone that prompts maternal behaviour and gets released in women while and after they give birth, also enhances their perceptions of cuteness.

Men don't experience the same guidance from their hormones, but that's not to say that their brains don't respond protectively when they view infants, whether their own or one in the psychologist's laboratory. In one study, men tended to express slightly less attraction to infant faces than women did, but there was no gender difference when it came to how hard men and women were prepared to 'work' (press a key – like a pigeon in a Skinner box) to look at infant faces, so on some measures men seem just as big suckers for cuteness as women are. Also, some men may feel inhibited in expressing a strong liking for pictures of babies simply because in Western society men aren't supposed to be interested in them (apart from their own). Overall, men and women may be rather similar in their susceptibility to cuteness even if its effects on their behaviour differ slightly (for example, brain studies have shown greater responses in the right brain in women and the left brain in men).[3]

Women's experience of babies, both short and long term, also affect their reactions to cuteness. Those who were brought up with no siblings or had a large age gap between themselves and their younger sibling (and therefore did not experience babies on a daily basis when they were growing up) tend to be less responsive. Also, context is key: in the short term, the features of one (unfamiliar) baby can affect the reaction to another. After seeing pictures of several not-so-cute babies in succession, women tend to rate a baby with extra-cute features more highly than they otherwise would, and vice versa.

It's testament to the power of cuteness that it works its magic on men and women alike, affects those who have not yet had children of their own as well as parents, and applies to unfamiliar babies as well as those known to the perceiver. It also transcends racial and ethnic differences. In one study conducted in Switzerland, women (all Caucasian) preferred cute African infants to less cute African babies to the same degree that they preferred cute infants of their own race. The men (also Caucasian) were actually more discriminating when asked to choose between the African babies, specifically when determining which they would give a toy to. None of the people tested would have had extensive experience of African infants and so presumably based their discrimination of their cuteness on instinctive reactions to the features of the 'baby pattern'. Famine relief charities, which frequently feature infants when appealing for funds, seem exquisitely aware of this phenomenon.[4]

Brain scans support the idea that cuteness matters in how we respond to children. Dozens of small areas of the brain respond to infant faces, some in the cerebral cortex and others in more primitive regions, such as the thalamus and cerebellum. Adult faces also trigger most of these, but many respond faster and/or more intensely to faces of infants. Initial facial recognition takes place in the fusiform face area, with the right fusiform gyrus being especially sensitive to cute features (see nearby illustration). Immediately on detection of a face, several areas in the parietal lobe of the cerebral cortex direct attention towards it. Then other areas of the brain generate the instantaneous empathetic reactions that an infant face engenders: warm feelings towards the infant get triggered in the reward and attachment systems, including the orbitofrontal cortex. In parents, the mentalizing network, which processes anthropomorphism, becomes active, as the mother or father tries to work out what the infant might be thinking. Finally, a variety of motor areas start firing, preparing

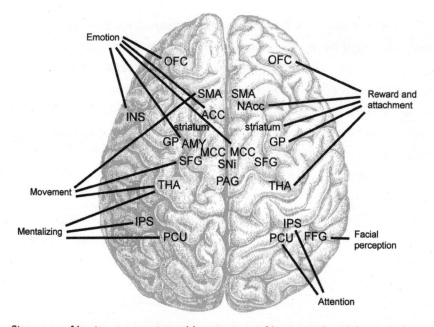

Emotion

OFC   OFC

SMA SMA   Reward and
NAcc   attachment

INS   ACC

striatum   striatum

GP AMY   GP

MCC MCC

SFG   SNi   SFG

THA   PAG   THA

Movement

Mentalizing   IPS   IPS

PCU   PCU   FFG — Facial
perception

Attention

Six types of brain areas activated by pictures of human infants' faces, divided by the function they are thought to perform. For abbreviations, see note.[6]

mothers (especially) not only to move towards their infants but also to speak to them.[5]

A few areas of the brain seem especially important in that they are particularly responsive to the cutest features. In addition to the fusiform gyrus, these include the left anterior cingulate cortex (empathy and emotion), precuneus (possibly a link to conscious thought) and the right nucleus accumbens, which is also an important target for oxytocin.

The complexity of these systems stems partly from the apparently hodgepodge way evolution constructed the human brain. Even so, the sheer number of parts of the brain that light up when a woman (or a man) sees an infant itself demonstrates the importance to our survival of responding appropriately. The apparent contradictions in many people's spontaneous emotional reactions to cuteness, which can combine feelings of euphoria with weeping, more usually associated with sadness, also reflects this complexity.[7]

More than just another word to apply to something we like, 'cute' describes a very specific set of biological responses to infants. We – men and women alike – possess an instinctive set of perceptions and reactions that drive us to pay attention to and care for our infants and children. The child's visual features trigger some of these; others (less well studied) rely on sounds that the infant makes or its clumsy movements. Some baby animals share these features, so we unsurprisingly use the same word when we recognize these similarities.

CUTENESS IS SO evocative that it has driven the depiction of animals in the media. Science may not have turned its attention to cuteness until the 1940s, but by then artists and advertisers knew the power of cute well. These 'natural experiments' show how, freed from the constraints of actual biology, we desire to make animals ever cuter, until some become total caricatures.

Cartoonist Rose O'Neill could have taken the specifications for her Kewpie characters from Konrad Lorenz's book, but they first appeared (in the *Ladies Home Journal*) in 1909, long before Lorenz published his list. Kewpies were baby-like creatures, usually naked but sexless, with tiny wings attached to their shoulders. They became so popular that O'Neill was at one time the highest-paid illustrator in the world. Children were the ostensible target for most of the stories she wrote about the Kewpies, but their potential to influence women could not have escaped her attention: as an early women's rights activist, in 1914 she permitted the use of a (beskirted) Kewpie to promote female suffrage (see nearby illustration).[8]

The Kewpies were baby cute from the outset, but another, more enduring cartoon character, Mickey Mouse, evolved to become cute, gradually losing his most animal-like characteristics. In his initial incarnation, in the movie *Steamboat Willie* (1928), Mickey was mischievous, even cruel at times, and despite being heavily anthropomorphized (wearing trousers, walking on two legs, and having hands), he was otherwise recognizably mouse-like. Over the following quarter-

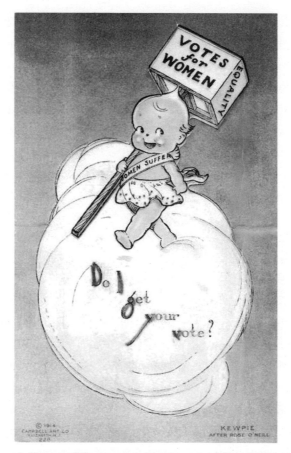

A Kewpie Elf – its cute features used on a 1914
postcard promoting female suffrage.

century he gradually morphed into the much more baby-like, in-offensive character that we know today. His appearance changed – somewhat haphazardly, suggesting that the Disney artists were not plotting a conscious trajectory – to become much more juvenile. His designers thickened his limbs, making them appear shorter, and en-larged his head and maximized the apparent size of his forehead by moving his ears to the back of his head; his eyes grew from 27 per cent to a whopping 48 per cent of the length of his head. In a nutshell, to sustain his public appeal, Mickey went from adult to infant, graphi-cally demonstrating the average consumer's preference for cuteness.[9]

Nowadays, other characters have eclipsed Mickey Mouse, most notably Hello Kitty, the ultimate manifestation of two-dimensional anthropomorphized cuteness. Instantly recognizable from her head alone, Hello Kitty ('real' name Kitty White, 'born' in London, but invented in Japan in 1974) requires just nineteen black lines to depict her almost total infantilism – round face (sans mouth), tiny nose, and widely spaced (though not large) eyes. The ageing demographic of her native land has prompted the globalization of Kitty, who appeals mainly to female children: the first overseas Kitty theme park opened in China in the summer of 2015.[10]

The teddy bear evolved into its current form along much the same path travelled by Mickey Mouse.[11] An exhibition of teddy bears in the Cambridge (England) Folk Museum in the mid-1980s provided an opportunity for two researchers at Cambridge University to test whether the evolution of Mickey Mouse also applied to the teddy bear – and it did. The earliest bear in the collection, dating from 1903, was naturalistic, to the extent of having a long snout (and wearing a muzzle, like a bear in a circus). Gradually, over the following half-century, the bears' heads became larger and rounder, and their foreheads became more prominent; though these changes are not quite the same as those Mickey Mouse underwent, they certainly support the idea that Lorenz's Kindchenschema drove their evolution.[12]

Of course teddy bears do not breed, so their evolution most likely reflected the preferences of the people buying them, many of whom would be women and therefore particularly susceptible to the attractions of the 'baby pattern.' Indeed, an infant ought not to prefer a toy that closely mimicked its own features, since the toy might then compete for its mother's attention (both my grandsons favour their 'rabbies' – plush toy rabbits with large soft ears but smaller-than-natural heads). Cuteness does not become a major player until children get older. At around the age of six, both boys and girls begin to show a preference for baby-like features. When asked to choose their favourite from a selection of adult- and baby-featured bears,

17 out of 24 four-year-olds chose an adult-featured bear, contrasting with the 20 out of 27 eight-year-olds who picked out a baby-featured one.[13]

Our preference for cuteness powerfully drives the features we prefer in representations of animals, be they soft toys or cartoon characters. Unconstrained by biological realities, creatures such as Mickey Mouse and Hello Kitty prove that given the choice, adult humans prefer the cute over the less cute. Real animals have less scope for change, since they must preserve basic functions – for example, as her forehead enlarged to the cutest possible proportions, Hello Kitty's mouth all but disappeared, and it's unclear, given the size of her head in proportion to her feet, how she avoids toppling over. Despite the constraints of reality, abundant evidence shows that cute features affect how people feel about living animals. Moreover, now that fewer and fewer cats and dogs work for their living, we can detect parallel changes in their anatomy, with increasing numbers of owners opting for the more 'baby-faced' breeds such as Persians and pugs.

ACCORDING TO ONE theory as to why we keep pets, their cute features seduce us (especially women) into taking care of them. In this concept, pets are little more than substitutes for human babies, triggering behaviour that we should, by rights, reserve for human infants. Indeed, Lorenz's original hypothesis proposed that cuteness transcends species boundaries: not only should mother cats find their kittens cute and therefore want to care for them, but so should we. Adult humans certainly do prefer certain cute features in, for example, dogs. Puppies induce feelings of tenderness to a far greater extent than even the friendliest of adult dogs. Subtle alterations to pictures of the heads of various types of dog have shown that most people prefer those with larger eyes, more space between them, and smaller jowls (the last-mentioned feature probably mimics the small chin of a human infant). Subtly increasing the size of the lower jaw and decreasing the height of the forehead can render cats in photos less cute.[14]

Both direct experience and cultural norms undoubtedly influence people's reactions to pets, so demonstrating that responses to cuteness are instinctive is not entirely straightforward. Perhaps children gradually learn how to react to defenceless mammals by imitating their parents rather than obeying the gradual emergence of instinctive responses to cute features. However, one almost-forgotten experiment conducted in Vienna in the 1950s produced a very similar pattern of results to that found for teddy bears, but with real animals. Anthropologist Paul Spindler examined children's naive reactions to kittens by recruiting his forty-eight subjects from an orphanage where they had lived since birth and where there were no animals. Unfortunately his previously testing them with a snake may have coloured their reactions to the kittens somewhat, but nevertheless the experiments were remarkably revealing of children's instinctive responses to cute, friendly animals.[15]

The children's reactions showed marked differences depending on how old they were. Many of the younger children – those up to about three years old – were initially nervous (some burst into tears and had to be consoled by their nurse, who was sitting nearby) but then treated the kitten much like a toy, reflecting their inexperience with animals. Those between three and eleven years old behaved quite differently: both boys and girls smiled, spontaneously talked to the kitten, and picked it up, often cradling it protectively in their arms like a human baby. Typically they laid the kitten on one forearm and stroked it with their free hand, bending their faces down to make eye contact and trying to converse with the kitten using baby talk (see nearby illustration). The girls would have had opportunities to practise such actions with their dolls, but the boys, presumably, less so, suggesting that their behaviour was largely instinctive, although even they would have observed nurses in the orphanage handling babies. Thus as children mature they not only start to prefer cute features (as shown by the teddy bear experiment) but react to them in a caring way similar to how human mothers treat their infants.

In adults, simple photographs of kittens or puppies can trigger caregiving behaviour. One study co-opted the classic US children's game Operation to test whether women become more cautious and patient after viewing images of cute animals. (The game consists of

Boys from an orphanage who knew little about animals
instinctively treated cats like babies (redrawn from
photographs in the original 1961 scientific paper).

using metal tweezers to remove small plastic objects – 'body parts'
such as bones – from an image of a human body. The body parts
are situated in confined spaces with a metal rim. Contact between
the tweezers and the rim triggers a loud buzzing sound, indicating
that the operation has damaged the 'patient'.) The women's performance improved – that is, they were much more careful with the
'patient' – after they looked at just three pictures showing a puppy

or a kitten as opposed to an adult cat or dog. Somewhat surprisingly, men showed the same effect, albeit to a slightly lesser extent. Thus, when people view cute animals, they not only feel tenderness towards them but also momentarily become more physically tender in general. Women who had seen eight pictures of cuteness-enhanced infants and toddlers – with slightly enlarged eyes, cheeks and foreheads and slightly reduced noses, lips and chins – demonstrated the same changes.[16]

Although the majority of studies have focused on animal faces, Lorenz's original conception of cuteness also encompassed other aspects of the body, particular a large head, short chubby limbs, and clumsy movements. Puppies and kittens embody all these and thus unsurprisingly get rated as cuter than their adult counterparts. However, many people consider adult dogs and cats cute in their own right. Cats, with their round faces, large foreheads, and big, widely spaced eyes, arguably fit the 'baby pattern' quite well, even though their limbs are no longer chubby, and their movements are anything but clumsy.

Some types of dog exhibit baby-like features. The appeal of breeds with flattened faces – the technical term is brachycephalic – must derive from their similarity to humans (to their detriment – see nearby box). On the other side of the body, so to speak, the short legs of the dachshund, corgi, basset hound and similar breeds (caused by a single mutation to a growth-factor gene) not only resemble those of puppies but give such dogs their characteristic waddle. Toy dogs can be picked up and carried as if they were human infants. Yet not all dogs are physically babyish, so if that quality were their only appeal, their owners would tend to reject them once they had left puppyhood behind.

What, then, makes a full-grown Great Dane or German shepherd appealing? Although not physically babyish once they grow up, breeds like these do remain playful – unconsciously mimicking the playfulness of small children – right into adulthood. Adult cats also continue to play (mainly with toys). Perhaps by playing, both cats and dogs continue to trigger our sense of their cuteness throughout their lives, even once they have lost some of their most babyish facial features.[17]

We should beware of automatically presuming that calling a dog cute can only mean 'baby cute'. Consumer researchers have identified a second dimension to cuteness that induces people to feel care*free*, not care*giving*. Advertisers use images of cute children and animals to promote all sorts of things – many in no way connected to children or animals. Market researchers have proposed the concept of 'whimsical cuteness' to explain a fascinating phenomenon: these images suppress common sense and induce people to become self-indulgent (that is, to want to buy things that they do not need). For example, in one study people took 30 per cent larger servings of ice cream when using a scoop with a handle shaped as a woman whose beehive hairdo formed the scoop, as compared to a plain but similarly coloured implement. They had rated the decorated scoop as whimsical, fun and playful but not as vulnerable or needing care, as they would had baby cuteness been triggered. Similarly, viewing a biscuit decorated with a cartoon lion's face led many people to prefer a more indulgent dish (rich, delicious, but fattening) over a healthy option for their meal that evening. However, if the biscuit was described as coming from 'The Kids' Cookie Shop', most picked the healthy option instead. The two types of cuteness appeared to have cancelled one another out: the feelings engendered by baby-pattern cuteness, induced by the name of the shop, inhibited the whimsical cuteness effects of the cartoon face.[18]

Thus when someone calls a dog cute, we can't immediately be sure whether he or she means baby cute or whimsical (fun) cute. If the dog is small, or flat-faced, or has stumpy legs, the person may mostly find it baby cute (if it's a pug – which has all three of these traits – then that's pretty much a foregone conclusion: see nearby box). If the dog is medium in size or large with a long nose and long legs, such as a greyhound, its appeal may also include fun cute – due to the person's experience that it's playful and fun to be with – although playful must also cross over into baby cute, since toddlers are playful.

Cuteness – a Pug's Perspective

People may find flat-faced dogs and cats especially appealing due to their resemblance to human infants, but these features can come at a huge cost to their welfare. The 'Peke-faced' variety of Persian cat, with its nose squashed between its eyes, is prone to breathing difficulties, eye disorders and malformed tear ducts, and a higher than normal proportion of its kittens are stillborn. Breeding of the British bulldog, once a proud (and athletic) symbol of the nation's reputation for pluck and determination, has favoured such an extreme head shape that most puppies can no longer travel down the birth canal and must be extracted by Caesarean section. Because of their deformed nasal passages, many have difficulty breathing when asleep and may force themselves to stay awake to avoid suffocating. This affliction stems from the excessive amount of soft tissue in the dog's airways: the mutation that causes the flat face affects the growth of the bones in the nose, not the tissue that the skull supports. Thus these dogs have undersize nostrils, excessive folds of tissue in the system that normally humidifies the air that passes through the nose, an oversize tongue, a soft palate and tonsils, and excessive folds on the larynx. As a result, once they reach adulthood few can breathe

A pug.

through their noses and find it increasingly difficult to breathe even through their mouths. As they struggle to inflate their lungs they must adopt a wide stance, with their elbows pulled away from their chests, which resemble the enlarged chest of a child with severe chronic asthma. Such dogs are virtually incapable of taking exercise, fully occupied as they are with just breathing. Eventually the larynx collapses under the strain, and the dog suffocates.

The pug has it even worse than the bulldog. A recent survey of dogs in the London area indicated that while any dog with a short muzzle could have difficulties with breathing, the worst-off breed was the pug, with around 90 per cent of dogs showing signs of an obstructed airway.[19]

The unfortunate pug's problems don't stop at breathing difficulties. Their extra-large eyes, reminiscent of a human infant's, are very susceptible to several kinds of problems, including inflamed eyelids and damage to the protruding cornea, resulting in partial or even total blindness in the affected eye. Worse – and this is not for the squeamish – a simple fall can knock the eye right out of its inadequately sized socket: one-eyed pugs are not uncommon. As if that were not enough, the excessive folds of skin just above the nose can harbour skin infections, and their hair can scratch the eyeball. Finally, there isn't enough room on the upper jaw for all the teeth, some of which turn sideways as they grow, predisposing the dog to painful gum disease later in life.

All these handicaps stem directly from owners' desire to possess the ultimate baby-faced dog. In the words of one distinguished veterinary behaviourist, 'I love pugs. That's why I don't own one.'

Furthermore, cuteness is not simply a trait that a pet has or doesn't have; it also mutates in the eye of the beholder. In a recent study conducted in Australia, dog owners shown pictures of other people's dogs consistently rated some as cuter than others. However, the owners not only rated their own dogs as cuter than anyone else did, but their ratings also bore no relationship to those of people who had only seen the dogs' photographs.[20] As owners establish a relationship with their dogs, it seems, they rate them as increasingly cute. Presumably their appearance becomes less important as other attributes become more so, especially how loving the dog turns out to be. Nevertheless, the initial impression of cuteness (or otherwise) seems not only more or less universal but also important in jump-starting the relationship between man and his best friend.

WE USE THE WORD 'cute' to describe both human infants and baby animals, and undoubtedly the two overlap a lot in how they make us feel, even though we know perfectly well that a 'fur baby' is not actually a human baby. Some studies suggest almost total overlap, as if cute animals simply hijack our instinctive emotional reactions to infants of our own species. Many people rate even adult cats' faces (let alone those of kittens) as cuter than those of (some) human infants. In another study, people shown pictures of cute puppies then rated human infants' faces as less cute than they otherwise would have, as if the puppies had satisfied their appetite for cuteness and the baby pictures were just too much of the same old thing.[21]

In contrast, several other studies have indicated that our responses to cute animals and to cute infants are distinct from each other. Pictures of human infants grab our attention faster than those of equally cute puppies and kittens, especially if presented to the left eye only and processed predominantly by the right side of the brain. (Mothers' preference for holding their babies on their left arm, allowing them to monitor the child's facial expressions with their left eye, shows a similar bias.) In another study, viewing faces of human infants elicited an instantaneous positive emotional reaction not seen with puppies or kittens (or adult humans). Magnetic resonance scans have identified several areas of the adult brain that respond strongly to human infants' faces but not to those of puppies or kittens. These include areas concerned with emotion (insula, medial cingulate cortex – see illustration on p. 186) and with preparing the brain for action (thalamus, supplementary motor area, fusiform gyrus). Although we may say that we find human and animal infants equally cute, it seems that our brains are subconsciously clear that they are not at all the same and prime us to behave differently towards them.[22]

While we may refer to pets as members of our family, the scientific evidence suggests that our brains do not respond to our adult dogs in precisely the same way we respond to our children. If pets are child substitutes, owners' affection for them should show up in the same brain areas as that for their children. In one study, brain scans of dog-owning mothers did show a great deal of similarity when they looked at pictures of their own dog or child. The amygdala, the peri-aqueductal grey, and the substantia nigra, all of which contain high concentrations of receptors for the hormones oxytocin, vasopressin and dopamine, lit up indiscriminately. However, while viewing the child also led to activity in a brain area also involved in affection-ate relationships with people (the ventral tegmental area/substantia nigra), viewing the dog did not. The affection that owners feel for their adult dogs is thus not identical to that which they feel for their children.[23]

The way mothers talk to their infants also differs subtly from how (female) owners talk to their dogs. Baby talk, or 'motherese', has been well studied and differs from adult speech in several distinctive ways. It is characterized by very short utterances, typically about four words long, two and a half times shorter than a typical statement made as part of a conversation between two adults. Mothers usually keep to the present tense when talking to their babies and make a low proportion of statements and a high proportion of exclamations and questions: 'Where's the doggie? Where's he gone? He's over there, isn't he!' They often repeat words, use many diminutives, over-articulate vowel sounds, and speak in a higher pitch than usual. (Men use some baby talk when addressing their babies (or dogs), though to a lesser degree than women.) Baby talk appears to be evolutionarily shaped, since babies pay special attention to motherese even when being spoken to in a language they have never heard before. Mothers do not expect their babies to understand every word they utter, just as surely as most owners do not expect their pets to understand them completely.[24]

Motherese is superficially similar to 'doggerel', as it has been called (reclaiming the original root of the Middle English word coined to describe bad poetry). The American novelist Booth

Tarkington, in *Seventeen* (1917), gives this example of an eighteen-year-old girl talking to her dog: 'pressing her cheek to Flopit's, she changed her tone. "Izzum's ickle heart a-beatin' so floppity! Um's own mumsy make ums all right, um's p'eshus Flopit".' In Japan, women talk to their dogs in much the same way as they do to children. They add the diminutive *chan* to a dog's name in the same way they would when addressing a young child (*Dale-chan*), or use the word *ko*, which denotes a young child (for example, *iiko* for 'good boy' or 'good girl'). They refer to areas of the dog's body using diminutive words, for example *okuchi* rather than *kuchi* for 'mouth', and speak of themselves and their husbands as the dog's mother (*okaasan*) and father (*otousan*).[25]

Although there have been few studies of cat owners' use of language with their pets, it seems to resemble doggerel. One study appeared to reveal some minor differences: women use even shorter utterances (most consisting of just two words) when speaking to their cats and a lower proportion of questions. However, the people involved had never met the cat before being asked to play with her for the short period during which their remarks were recorded. Owners conversing with their pets in the privacy of their own homes might well be more expansive.[26]

Doggerel and baby talk also differ in that mothers constantly point out the names of things to their infants: 'This is a ball. That is red.' Owners don't usually do this with their dogs in informal situations, although they may include similar utterances when attempting to train or generally instruct their dogs. More tellingly, doggerel also differs subtly from motherese in its vowel sounds, which people exaggerate less when talking to dogs. It's known that the infants of mothers who over-articulate their vowel sounds become better able to distinguish between the consonants that the vowels separate, so, like the use of statements, this attribute of motherese seems to have a function in teaching the child to speak. Since few dog owners expect their dogs to learn to talk back to them (the 'talking' dogs on YouTube notwithstanding), it makes sense that they would drop this particular aspect of motherese. Owners of parrots, however, do expect that their pets will pick up fragments of human speech, and when people talk to their parrots, some (not all) do over-articulate their vowels, as if they were talking to a baby.[27]

Modes of speech derived from motherese are used in a wide variety of contexts, of which speaking to pets is just one (perhaps two, if we include parrots, or even three, if cattish turns out to be subtly different from doggerel). Other common recipients include lovers, foreigners, infirm elderly people, mentally challenged adults, and robots. Romantic partners use baby talk mutually as a way of simultaneously expressing affection, commitment and dependence, and although the words used may differ, the exchange of sappy terms of endearment shares striking similarities with doggerel. Less similar are the versions used when the speaker believes that the recipient has difficulty understanding what is being said, either because of a language barrier or a perception of diminished competence; these generally have a less affectionate tone than that used when speaking to babies, lovers or pets. Such a distinction emerged from a direct comparison between the verbal instructions issued to a Boston terrier and an AIBO robot as the subjects in an experiment directed them through a complex maze. They issued very similar instructions but used a noticeably higher-pitched – more affectionate – tone of voice for the dog than for the robot.[28]

The way people talk to their pets may therefore reflect more generally their affection for the animal rather than their conception of it as their 'baby'. Although motherese almost certainly evolved as a way of reassuring babies and teaching them to talk, we have apparently adapted it to a wide range of other situations where normal adult-to-adult speech seems inappropriate. Each new function entails an alteration, generally a loss, of one feature of its original, infant-directed form. The words change when romantic partners converse, the articulation shifts when dogs and foreigners are involved, and the affectionate tone drops away for foreigners and robots.

There can be little doubt that our instinctive reaction to cuteness is a part of our mammalian heritage that ensured the survival of our offspring, going back millions of years to our hominid past and probably further than that. Yet the ways in which our pets trigger these mechanisms differ subtly from those that human infants initiate. The sight of cute animals activates slightly different areas of the brain as compared to the sight of an infant, and our verbal response also differs subtly. The many similarities should not lead us automatically to conclude that pets are simply exploiting a set of responses that we ought by rights to reserve for our own species. Natural selection favours minimal changes rather than wholesale reinvention. If at some time in our past it became advantageous to evolve mechanisms for recognizing and then taking care of some kinds of animals, then we would expect to see the minimum number of changes possible to an already existing response – in this case the own-offspring-directed cute response. Part of the appeal of pets is borrowed from the appeal of our infants, but it is adapted, not plagiarized.

CUTENESS IS AN essential part of the initial attraction that people feel for puppies and kittens, cats in general (probably), and many kinds of dog. (It is probably much less important, and may even not be a factor, in many other human–animal relationships, such as those between horses and riders or herpetoculturists and their snakes, however affectionate those people consider their bond with their animals to be.) Although cuteness is clearly a property of both human infants and some animals, and stimulates caregiving towards both, most people do not react to these two sets of stimuli in precisely the same way, as shown by the differences in brain activations and in the type of speech used to address them.

Cuteness (in the sense of baby cute) is unlikely to be enough to sustain a real relationship, even though it may continue to play a part in the feelings that a child has for his or her favourite teddy bear. The initial appeal of a cute face may account for much of the popularity of cats (and pugs) on the Internet, but people's responsiveness to the cuteness of a specific face – animal or human – wanes with time, so it is unlikely to account for the durability of owners' bonds with their pet dogs and cats, which can last a decade or longer. (Confusing the picture somewhat, long-standing pet owners may casually use the word 'cute' when putting their affection for their dog or cat into words.) Both parties must get something out of the relationship if it is to last. Pet dogs value social contact with people above all else, cats the consistency of their surroundings and the protection that their owners provide. So what do humans get out of the interaction over the long run? Perhaps a sense of enjoyment, rather than just cuteness, lies behind the affection that owners feel for their pets. Pets make us happy in a way that other possessions do not.

CHAPTER 8

Friends Forever – but
What's the Pay-off?

Most pet owners expect their dog or cat to be a friend for life (its life, that is). People widely view abandoning a pet for selfish reasons as somewhat shameful, though nowhere near as contemptible as abandoning a child. This contrasts starkly with what we consider acceptable behaviour towards the gadgets that today's pet owners likely give at least as much time to as they do to their animal companions. People buy and discard mobile phones and tablets seemingly on a whim. The United Kingdom has considerably more mobile phones than people (users, on average, have two phones they don't use, in addition to the one they do), but at least each phone has had in the neighbourhood of two and a half years of use; in the United States that number declines to a meagre one and a half years.[1] Although pets are possessions, they are regarded much more like people.

Reliable statistics on the proportion of dogs and cats abandoned by their owners at some point in their lifetimes are hard to come by – especially because many transfers of ownership are private affairs and do not involve data-gathering organizations (such as one of the major humane societies). Cats are somewhat prone to migrating spontaneously from one home to another – usually due to social incompatibilities rather than the oft-blamed quality of food on offer. Many dogs end up in shelters because their owners find themselves unable to cope. Nevertheless, nowadays we generally believe that both species – dogs especially – can and should be friends for life.

The joys of pet ownership must be both powerful and robust to cancel out the downsides. One 2011 survey of 2,000 UK dog own-ers revealed that the average dog triggers over 150 family arguments each year, or around 2,000 during each dog's lifetime. Holidays of-fered the biggest bone of contention: Should the dog be allowed to come along? If not, who should care for it? Who walks the dog is also a source of friction: although three-quarters of owners say they like walking their dogs, many try and fail to get other family members in-volved. Another source of disputes entails accusations that one fam-ily member is too soft on the dog, allowing it upstairs or even onto the bed; in 17 per cent of households this indulgence had forced one family member to sleep in a spare room, leaving the dog to occupy the space he (or she) had vacated.[2]

In order of occurrence, the top twenty doggy disputes had to do with the following:

1. What to do with the dog when going away on holiday/for the weekend
2. The fact the dog hasn't been walked/who should walk it
3. Whether the dog should be allowed on the bed
4. Whether the dog should be allowed upstairs
5. Who should clean up the mess in the back garden
6. Whether one of the owners is being too harsh on the dog
7. Whether one of the owners should let the dog onto the sofa
8. How much one of the owners has spent on the dog
9. Who should train the dog and how
10. Whether to allow the dog to be fed from the dining table
11. Who should babysit/look after the dog
12. Who should groom the dog
13. Household damage caused by the dog
14. Who chose to buy the dog in the first place
15. Who should clear up the mess when the dog urinates or defecates on the carpet
16. Who should clear up the mess when the dog is sick on the carpet

17. One of the owners insists on treating the dog like a human
18. One of the owners lets the dog into a room of the house it is not normally allowed into
19. The kid's toys have been eaten
20. Shoes have been chewed

For some owners, such everyday annoyances may pale into insignificance if their dog becomes aggressive or cannot be left on its own because it 'misbehaves' (most likely due to the stress of separation, which is much more common than most owners imagine). Some, seemingly unaware that expert help is available, come to accept such problems as part and parcel of owning a dog. Couples stop entertaining at home because their dog growls and snaps at strangers; a person lives alone because when she returns from work, she has to clean up the faeces in the room where she has confined her beloved dog, unaware that its behaviour reflects the distress it feels when she leaves each morning.[3]

Bothersome behaviour is not the only downside of dog ownership. The past quarter-century has seen the advent of a new burden for dog owners: disposing of their dog's faeces from public spaces. In the 1950s, dogs were supposed not to defecate in a few places – railway carriages, for example, although the station platform was fair game. In the West, the joy of taking the dog for a walk has now been sullied, for some, by the need to retrieve, and then dispose of, its droppings. In the words of broadcaster Robert Hanks, 'For any dog owner, shit is a major issue . . . [H]owever often you pick up the shit, however much you love your dog, the hideousness of the task does not diminish. However deep the lesson of good citizenship has sunk, to walk down the street with a sack of faeces in your hand, looking for that elusive next bin, is a humiliation.' Yet despite this additional responsibility, the popularity of dog ownership shows few signs of diminishing.[4]

There are fewer recorded examples of how cats might disrupt family life – probably because they are less demanding than dogs. But their strong attachment to place may impose burdens on an owner and thereby influence decisions over living arrangements: 'I think

sometimes, it would be more convenient if I had a little flat somewhere but this is home,' said one cat owner in South Wales. 'And
I've got a cat and he wouldn't like being in a flat because we did have
him in a flat when we were over there when the house was being re-
done and he wasn't happy at all, bless him.' More mundane inconveniences include the 10 per cent or so of pet cats that regularly urinate
or defecate in corners of the house or, worse, on beds (according to
a survey I conducted in southern England). (Usually, such behaviour
reflects the cat's fear of being terrorized by other cats. In the same
survey, we found that about one cat in eight was aggressive towards
people, usually visitors to the home, so this problem, although more
serious in dogs, is not confined to the latter species).[5]

Pet keeping also has social consequences – both for the owner
and for innocent bystanders. The benign image of the family dog as
cure-all and companion is at odds with its origins as a pack-hunting
carnivore: many inadequately socialized pet dogs can resort to aggression in stressful situations. Canine aggression is and probably always
has been a major problem worldwide because of the injuries that dogs
can inflict. In recent decades we've become far less tolerant of these
incidents, the less serious of which used to be shrugged off as just another of life's little hazards. Hospital admissions due to dog bites have
increased 75 per cent between 2005 and 2015, but this has more to do
with how seriously we take these episodes than with an actual rise in
instances. In the United States each year, dogs bite between 4 million and 5 million people; of these, twenty to thirty, mainly children,
die as a result.[6]

Dog bites aren't the only problem. Dog noise and mess are a frequent source of disputes between neighbours. My daughter recently
had to move house because the incessant barking of the 'home-
alone' dog next door prevented her infant son from sleeping during
the day. Dogs can also cause traffic accidents by running out into
the road or distracting their owners while they (the owner, not the
dog) are driving. And sometimes pets cause damage just by being
there. Each year in the United States up to 100,000 people require
medical attention – in 25,000 instances for bone fractures – after

tripping over their dog or cat. Cats seem to pose a special hazard for the over-75, and one wonders if many more such incidents go unreported by elderly people who worry that they might lose their pet if it were ruled a danger to them.[7]

Given the many potential pitfalls of owning a pet, owners must have strong countervailing reasons for continuing to care for and about an animal that is causing problems, either for them or for their neighbours. Feelings of social obligation can play some part. An elderly person may persist in keeping a pet given to him or her unexpectedly by a loved one, who hoped it might provide companionship and the supposed health benefits; fear of offending the giver may inhibit the recipient from finding a more suitable home for the animal. Empty-nesters may keep a pet that used to belong to a child who has left home, despite not having a particularly strong attachment to it; widows and widowers may keep the pet shared with a deceased partner despite having secretly been less than fond of the animal themselves. Owners of any age may feel ashamed of disposing of a 'member of the family', however intractable their problems with it.

Nowhere is the power of social obligation and peer pressure more apparent than in the handheld 'virtual pets' so beloved by Japanese teenagers. The first of these, the Tamagotchi, appeared on the market in 1996 and became an instant craze, with the $10 toys changing hands for up to $400 for a rare discontinued model; by 2010, over 76 million had been sold worldwide. Tamagotchi are egg-shaped key rings with a tiny screen that displays a stylized animal, allegedly an alien species (see nearby illustration). Activation of the toy causes an 'egg' to hatch, which players must then 'raise' by pressing the three or so buttons on the face of the toy. Each pet requires feeding, playing with, cleaning and disciplining. For example, it must be housetrained: droppings appear around the edge of the screen, and the pet will become 'sick' if players do not clear these away. Shortly before the pet is due to defecate, it pulls a face, and 'stink' lines appear around it; at this point the player can activate the toilet icon, and successful repetition will result in the pet becoming permanently toilet trained. A skull icon appears on the screen to indicate 'sickness' – for example, caused by the owner's under- or over-feeding or failing to pick up droppings – and if no 'medicine' is administered, the pet can die. Unlike a human infant or a live pet, the pocket pet provides little emotional feedback in terms of cues from facial expressions or behaviour, few cries of either delight or distress: the relationship centres on the owner's responses to the 'pet's' arbitrary and unpredictable demands. To the outsider, this may look like all work and no fun, but owners show surprising dedication to their miniature electronic parasites.

Simple peer pressure may suffice to induce Tamagotchi owners to take great care that these pocket pets 'survive'. Failure to take proper care of what, on the surface, amounts to little more than a screen, a microprocessor and a battery can have social repercussions among the owner's peer group. In the words of one eight-year-old

A pocket pet.

girl, 'Mommy, mommy! Angela let her baby die! She didn't check on her or feed her, and in the morning she was dead!'[8]

Avoidance of shame is not the only factor spurring owners to continue to care for their pocket pets. The thrall that such toys exert appears to derive from their inventors' exploitation of simple anthropomorphism. They are programed to appear unpredictable and autonomous (they are also cute, but in the whimsical sense, so this cannot account for more than a fraction of their appeal). They make demands on their owners, just as a real pet would: the original models 'decided' when they would make those demands, although following objections from teachers, later models incorporated a pause function so that their owners could focus on their schoolwork. They 'grow' from infancy to adulthood, requiring regular nurture from their – predominantly female – owners (for whom they were originally designed; apocryphally, those boys who buy Tamagotchi focus more on disciplining than nurturing them).[9]

Anthropomorphism, whimsical cuteness, and social obligation thus work together to sustain a child's interest in her pocket pet. When she temporarily tires of responding to its every whim, the thought of her classmates' censure if she lets it 'die' will spur her on. The same pressures, albeit magnified, apply when the pet is a living animal. Yet, however demanding pocket 'pets' may seem, their needs pale to insignificance in comparison with the amount of care that a real pet needs across its whole lifetime. Because looking after a dog or a cat can be a chore, the relationship must have upsides that more than justify the effort put in, at least for the majority of owners. Thanks to the scientific search for what might lie behind the supposed (and probably illusory) life-extending properties of pet keeping, researchers have quite thoroughly investigated the emotional rewards of interacting with dogs.

GIVEN ALL THE HASSLES, obligations, and expense incurred by pet ownership, there must be a plus side. On balance, pets make us happy. We don't merely hang onto them from a sense of obligation, although we may at times feel as though we do if the animal has done something particularly disgraceful. Why some things make us happy and others don't is notoriously difficult to pin down, but the way our body chemistry changes with our mood offers some clues. Researchers have tracked in pet owners quite significant changes in the levels of several hormones immediately after friendly interactions with their animals. These findings may go a long way to explaining how the bond between owner and pet gets reinforced once the immediate attraction has begun to wane.

A substantial body of work indicates that interacting with a real pet can have a calming influence (provided, naturally, that the pet is calm itself – attempting to stroke a fractious cat or separating two fighting dogs will never be relaxing). Direct physical contact with the animal itself seems important, leading to a reduction in the stress hormones epinephrine, norepinephrine and cortisol.[10]

One group of people almost guaranteed to feel anxious are undergraduate students during examination time. Over 1,000 universities around the world have put so-called animal visitation programs into place to help students calm themselves. For example, in May 2015 my own university (Bristol) got together with the Guide Dogs charity to offer stressed students the opportunity to relax in a 'puppy room': the 600 available quarter-hour slots filled up rapidly. Perhaps surprisingly, the sessions were almost as popular with men as with women. Research supports the students' enthusiasm: in one study, after as little as seven minutes' interaction with a friendly dog, students reported significantly reduced anxiety and greater contentment. Simply viewing a slideshow of pictures of the same dog for the same length of time had no effect on their mood or agitation. Moreover, although those students who already had experience of and liked dogs expected to get more from their session, past experience had no bearing on how much benefit they reported afterward. Physical contact and interaction with a friendly dog does seem to be genuinely calming in the short term; despite the proliferation of cat cafés around the world, we don't yet know whether an equivalent session with a cat has the same effect, but it seems plausible.[11]

While rewarding in its own right, the dogs' calming effect on students may also generate longer-term changes in attitude. Those students who responded positively to petting the dog will remember that experience and both subconsciously and consciously try to repeat it in the future – who knows, it might even prompt them to join the next generation of pet owners.

So WHAT MAKES people so happy when they play with animals? Does the simple repetition of these joys accumulate to produce the bond between owner and pet?

Several highly complex but powerful systems in the human brain guide us to repeat actions that have made us feel good – especially if those consequences are somewhat unpredictable (hence the lure of the slot machine). Indeed, such feelings can become goals in their own right (see nearby illustration). But the several neurohormones involved, the best known being dopamine and serotonin, get released

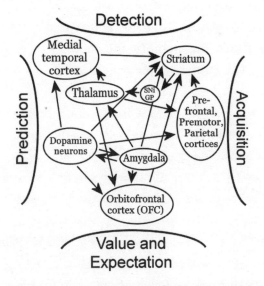

How the brain detects rewards and predicts where and when they are likely to recur, and how they might be acquired.[12]

in all manner of situations we find pleasant, not specifically when we interact with friendly animals. Other mechanisms must explain the affection we feel for our pets.

Physical contact with other living beings can produce changes in levels of another hormone, oxytocin, and researchers have made much of this recently in connection with our relationships with pets.[13] Historically, oxytocin was thought to have but one role in the body: initiating contractions during childbirth (intravenous administration of synthetic oxytocin is still used to induce labour) and then stimulating the mother to produce milk for her offspring. Natural oxytocin is manufactured both in the brain (mainly in the hypothalamus) and in a wide range of the body's organs, including not only the uterus and placenta but also the heart, the thymus (part of the immune system), the gut, and, despite its usual association with childbirth, the testes. Over the past two decades or so, research has found that oxytocin produces many psychological effects, helping to build not just the relationship between a mother and her newborn but also bonds between adults, between children, and, it has been suggested, between owners and their pets. Within the brain, nerves with oxytocin receptors make contact with several areas important for memory and feelings of reward and emotion, including the hippocampus, the nucleus accumbens and the amygdala.[14]

As with other hormones that have multiple functions, working out oxytocin's role in the human body at any particular moment is hard – partly because oxytocin in the brain and oxytocin circulating in the blood don't track one another precisely. Obviously for psychological effects we must scrutinize the levels in the brain most closely, but we can only do so by sampling the cerebrospinal fluid – an invasive procedure likely to have psychological effects of its own, which may in turn interfere with the very processes being measured. As an alternative, we can increase levels of oxytocin in the brain by administering the hormone in a nasal spray: this oxytocin apparently can slip between the olfactory receptors in the nose and up the sheath that encircles the olfactory nerve into the cerebral fluid. This takes about ten minutes, after which levels of oxytocin in the brain stay high for between one and two hours. Such doses have powerful – if usually short-lived – effects on people's behaviour and feelings, including appearing to build trust and enhance empathy and to lower anxiety and stress. Brain scans following administration of oxytocin show a reduction in the activity of the amygdalae and their connections to the sympathetic nervous system (which is responsible for the fight-or-flight response).[15]

According to one theory, oxytocin's anxiety-inhibiting properties are the key to understanding its several other effects. The human brain has many ways of assessing trustworthiness, but back in our primate past it was probably sensible to distrust everyone and allow only a few familiar individuals to come close – especially when a vulnerable infant was involved. Thus oxytocin, originally simply a trigger of birth and maternal behaviour, evolved an additional function: tamping down fear that physical contact with another member of the same species might otherwise trigger. Accordingly, it gets released into our brains when we receive signals that denote affection, such as eye contact, gentle touch (hence the effectiveness of massage in releasing oxytocin), and vocal expressions of fondness. The surge of oxytocin then induces us to mirror the same or similar behaviour, resulting in a cascade of mutual affection. In humans, this process plays a part in forming and reinforcing the bond between mother and baby, as well as in the formation of friendships between children and in the 'falling in love' phase of romantic relationships. The latter especially gave oxytocin its popular reputation as the 'love hormone' and led to its commercially availability as a nasal spray claiming to reduce stress, decrease anxiety, enhance orgasms and improve sleep. Some psychologists are also hopeful that oxytocin may provide a cure for autism, social anxiety disorder and schizophrenia.

Oxytocin does not, however, make us uniformly more trusting. Indeed, in some scenarios, high levels seem to generate less trust, not more: for example, breast-feeding mothers with high levels of circulating oxytocin can be more suspicious of people they don't know than their peers who have elected to bottle-feed their infants. Viewing cute babies raises oxytocin levels but, in addition to enhancing a mother's feelings of love, also induces protective instincts – making her potentially more aggressive if she perceives a threat to her infant. Women whose close relationships are foundering tend to have higher than usual levels of oxytocin in their blood, alongside high levels of the stress hormone cortisol, which seems at odds with oxytocin's reputation as a universal stress buster. Nasally administered oxytocin can cause men to become more trusting of people they know but less trusting of those they don't. Thus, saying that oxytocin promotes trust is an oversimplification; rather, it appears to change the priorities we give to different kinds of social information, lowering our guard towards those we already have reason to trust, while raising it against those we don't. This would explain the somewhat complex relationship between oxytocin and stress hormones, which depends on who's around when the oxytocin is released.[16]

'Dogs hijack the human bonding pathway': remarkably, this 2015 headline appeared not in a tabloid newspaper but in the number one science journal in the world, published by the American Association for the Advancement of Science.[17] It referred to a study conducted in Japan that purported to highlight a possible role for oxytocin in the domestication of the wolf. Researchers found a small increase in urinary oxytocin (not generally considered a particularly accurate measure) in dog owners after they had stroked their dogs and gazed into their eyes for half an hour. Other, earlier studies had also shown modest increases in oxytocin (measured in blood) after similar interactions, but mainly in women; men's oxytocin may actually fall. Unfortunately, no studies involving interactions with dogs have measured cerebrospinal oxytocin, which should more accurately reflect what's going on in the owner's brain.[18]

Physical contact – whether with infants or between supportive couples – boosts oxytocin, so the fact that stroking a cat or dog has the same effect comes as no surprise. Some have suggested that the long-term benefits of oxytocin for cardiovascular health may help explain why couples tend to live longer than singles. At one time, scientists hoped that the oxytocin released though repeated contacts with pets would explain why dogs (at least) seemed to improve human health. But since we now know pet keeping doesn't usually help people live longer, any effects oxytocin might have on owners' feelings for their pets don't seem to affect their health – or perhaps they do so only in the very short term, cancelled out by all the times when the dog misbehaves. Overall, despite all the hoo-ha about oxytocin's function in the formation of relationships, both between people and with pets, its effects are too inconsistent to account entirely for the pleasure we get from stroking an affectionate dog or cat. But for all the attention science has directed at oxytocin, it may have overlooked the role of another powerful class of bonding hormones: the endorphins. These two hormones play different roles in relationships between people: in general, the initial stages of a relationship – a mother's bond with her newborn, couples falling in love – seems to involve oxytocin more, while endorphins have a longer-term function. Endorphins can get released during physical contact, so perhaps stroking a pet affects us in another, possibly even more important way.

OWNERS LOVE TO stroke their pets – with cats, this is probably the most pleasurable type of interaction – so understanding how touch makes us feel must be key to grasping how the owner maintains a bond with a pet. The science examining touch has lagged behind that investigating our other senses, excluding our interactions with infants and sexual partners, and unfortunately most studies have looked at changes in the person touched, not the person doing the touching. Enough research exists, however, to suggest why we keep doing it and find it pleasant in its own right.

Our perception of touch has two clear dimensions. On one level, touch communicates information: where an object is and its texture. On another level, it's also emotionally rewarding. Even a light touch on the shoulder, from the right person, can affect both mood and behaviour. More protracted touch – hand-holding, for example – produces a measurable reduction in anxiety.

The texture of the entity touched affects people's state of mind. When we're feeling uncertain, we instinctively reach out for soft objects and 'worry' them between our fingers, which genuinely helps us to cope. Most of us find smooth skin – say, the palm of someone else's hand – less pleasant to touch than the hairier skin on the top of that person's forearm. And the human mind seems to react differently depending on whether someone touches us in hairy or smooth areas. Hairy skin has receptors that convey pleasant sensations to the brain of the person being touched; smooth skin doesn't.[19]

Presumably these and other consequences that arise from touching and being touched have their origins in our primate past, but the difference of our skin from that of our distant ancestors (and our habit of wearing clothes) hampers comparisons. Touch in general, and mutual grooming in particular, plays a very strong role in maintaining harmonious relationships within troops of monkeys. For example, baboons may spend up to one-sixth of their waking hours grooming one another; a few minutes a day would suffice for purely hygienic purposes, so the remainder must have social significance. Perhaps we find stroking our pets so pleasurable because it taps into some half-remembered primate instinct that we can no longer satisfy entirely by touching each other. Support for this idea comes from a study in which subjects thought they were helping with a 'consumer evaluation' of a plush-covered teddy bear: researchers allowed some to handle the toy and told the others to leave the toy on a table, out of reach. Before issuing these instructions, the researchers gave some of the participants (fake) bad news: they were to be excluded from a subsequent team exercise that other members of the group would be participating in. Touching the bear seemed to alleviate some of their unhappiness: those allowed to touch the bear were much more sociable afterward than those told not to handle it; apparently the pleasant sensation of touch had entirely changed their mood.[20]

Context profoundly affects how we experience the sensation of touch, especially our beliefs about whom we are touching or who is touching us. If we're shown a picture of an angry face – even when we're told that the person actually touching us is not at all angry – the touch feels less pleasant. If a heterosexual man is told that another man is stroking his arm – even though a woman is actually touching him – he will rate that touch as somewhat unpleasant. If he's recently been given a dose of oxytocin spray, then his ratings of the 'man's' touch will remain unaffected, but his ratings will increase if he believes that the touch has come from a woman. The powerful effect of context presumably accounts for why oxytocin sometimes increases in the blood when a person touches us and sometimes doesn't. Despite the absence of experiments, we can safely assume that the variable extent to which oxytocin rises in someone who's just stroked a dog depends on how that person feels about the dog in question.[21]

OXYTOCIN MAY BE important in the early stages of close relationships, and of course in the bond between mother and infant, but the smooth running of the complex long-term alliances that we form appears to require at least one other hormone system. According to a theory that's gaining increasing traction, endorphins, rather than oxytocin, may play the more critical role in maintaining long-term relationships in primates (so also in ourselves).

As with oxytocin, the primary function of endorphins is fundamental: first and foremost, they regulate the perception of pain, both physical and emotional – and opioid narcotics such as heroin target their receptors. Beta endorphin is largely responsible not only for raising athletes' pain thresholds while they're exercising but also for producing the simultaneous feeling of euphoria (aka 'runner's high'). Thus, in addition to their basic function, in humans endorphins play a major role both in subconscious reward systems and in conscious feelings of pleasure. Natural endorphins can become addictive, just as heroin can: if an anticipated reward suddenly becomes unattainable – say, when a romantically attached couple splits up – the feelings of loss can equate to withdrawal from opioid addiction.[22]

Endorphins play a major role in the tactile aspects of social relationships among primates, specifically increasing during grooming, tactile play, and huddling. These exchanges of physical contact comprise the glue that sustains long-term relationships in bands of the more socially sophisticated monkeys, buffering against the more immediate effects of other hormones, such as testosterone and prolactin, which tend to dominate the behaviour of lesser mammals.

Although our own social connections are more complex than those of the other primates, endorphins also seem to play a major part in our close relationships. Investigating these hormones in humans is not straightforward, so our knowledge of their workings remains incomplete. As with oxytocin, levels of endorphins in the brain (where their important social effects occur) don't track those in our bloodstream (which are easier to measure). Also, due to concerns about the after-effects, we cannot administer willy-nilly to humans drugs that mimic natural endorphins, like morphine, or block their effects. Nevertheless, a person who feels isolated will try to connect with other people, and successfully doing so, particularly if this leads to physical contact, will produce a surge of endorphins, which calms and relaxes the individual and at the same time tamps down feelings of anxiety. Heroin users can (temporarily) achieve similar subjective feelings from their fix, which explains why they tend not to be sociable people, apart from having pragmatic relationships with other narcotics users, sustained by their shared lifestyle.[23]

Because we humans touch one another far less often than our primate relatives do, considerable speculation concerns what act might have replaced mutual grooming as a releaser of endorphins in our own social behaviour. Researchers widely tout three candidates: music, dance and laughter, all of which can cause endorphin levels to rise. Pets are unlikely to bond with us through music – cats, for one, are pretty much tone deaf – and while some dog shows now include competitive dancing with dogs, it's not a regular feature of dog–owner relationships.[24] Laughter with (or at?) dogs, however, is commonplace.

Laughter is a primitive form of communication that almost certainly pre-dates speech. It acts as both an invitation to playful interaction and a bonding mechanism: its ability to raise pain thresholds, a well-known effect of endorphins, suggests a possible link between laughter and endorphin release in the brain. Laughter has two common triggers. One is cognitive: we laugh when we are surprised but don't feel threatened. The other is social: people who are fond of one another are most likely to answer laughter with laughter. Spontaneous laughter generates a release of beta endorphin, which in turn leads to feelings of well-being.[25]

We laugh at our pets when they do something unexpected. This applies especially to dogs. In fact, in one study cat owners laughed less overall than people who owned no pets, but this could just be a quirk of the cat owners surveyed rather than an effect of the cat. 'Funny cat' videos are of course massively popular on the Internet, so cats certainly can make people laugh. Dog owners laugh frequently when playing with their dogs, usually as a reaction to an unexpected action by the animal. Sometimes they laugh when the dog spontaneously stops playing; sometimes, when the game takes an unexpected turn (they also laugh, presumably from embarrassment, when the dog breaks off the game to pee). Nevertheless, a surge in endorphins may accompany such laughter and thereby help to cement the bond between owner and dog (so far, no one has investigated changes in people's hormones in such scenarios).

The benefits of laughter accrue only to the human though – not to the dog. Dogs don't seem to understand their owner's laughter. In the words of one study, 'What dogs make of our laughter, if anything, is a mystery.' Thus laughter in this context seems more a spontaneous expression of human instinct than a genuine piece of communication, at least not with the dog. It could be intended for any humans watching the game, although dog owners do report laughing when only their dog is there to hear them.[26] That the dog is a living being potentially capable of responding does seem to be important, however, since people rarely laugh when interacting with a robot dog.

We humans may have largely abandoned mutual grooming as a way of bonding with one another, but we still enjoy the sensation of touch, especially with our pets. Crucially, one study has shown that stroking a friendly dog can lead to a rapid increase in beta endorphin, in addition to the expected increase in oxytocin. Since the researchers measured these hormones in blood plasma, we cannot say precisely how much the level of each had changed in subjects' brains, but several of the owners' social reward systems were apparently activated, not just oxytocin. The increase in beta endorphin, especially, would lead to relaxation and reduced anxiety. Repeated day after day, such experiences enhance the bond between owner and dog in a way remarkably redolent of how our chimpanzee cousins cement their relationships. We might logically assume that the same applies to cat owners, though we lack studies showing as much.[27]

PERHAPS ACTING TOGETHER, the endorphin and oxytocin systems appear to provide the core reward that people obtain from stroking and patting their household pets. If the dynamics are the same as for relationships between humans, oxytocin may be more important during the relationship's initial stages – some of it triggered by the features of the pet that the new owner finds cutest – while beta endorphin becomes more important as shared experiences cement the bond. Although to date published studies have examined only dogs, we could regard the teddy bear, as described above, as a surrogate furry 'pet', and so such benefits should extend to cats. Thus, the gratification we obtain from general physical contact from our pets, particularly stroking (see nearby illustration), has a physiological basis. Furthermore, cats and smaller breeds of dog may allow us to express another of our instinctive primate bonding mechanisms: huddling together with another warm body. It's probably no coincidence that owners get so much satisfaction from cats that doze on their laps or dogs that lie across their feet (despite the eventual resultant cramp!).

Everyone in the household has access to the rewards of amusement and warm physical contact, but the costs of pet keeping – in both time and money – often fall less equally. However, the act of caring for a pet may be rewarding in itself, which perhaps contributes to the popularity of 'pocket pets', which are neither warm and furry nor particularly amusing. Although we know little about the mechanisms involved when the target is an animal, we can reasonably suppose them to be analogous with those activated in mothers when

Stroking the warm fur of a dog may tap into our primate
instinct to build friendships through grooming.

attending to their infants – most of them say that they find providing
this care highly rewarding. The underlying hormonal network comprises a specific pathway in the limbic system of the brain, which responds to a surge of dopamine in turn released by oxytocin. Repeated
activation of this system strengthens the bond between mother and
baby. It's entirely possible that something similar occurs when owners feed or groom their pets.

DESPITE ALL THE EFFORT – and cash – that owners devote to their
pets, most try to maintain a relationship with their animal companion for the whole of its life. We have not fully elucidated the complex rewards that motivate this seemingly altruistic behaviour, but
they likely extend far beyond the simple attribution of the bond to

any single hormone, particularly oxytocin. Other neurohormones al-most certainly play a part, as they do in mother–infant bonding in most animals more sophisticated than a vole: beta endorphin and do-pamine are the most likely candidates, but science has never mapped any hormone precisely onto any aspect of human behaviour, so it may turn out that several others are involved as well. The means whereby our pets make us happy remain somewhat obscure.

Studies have scarcely looked at relationships between men and their pets, certainly by comparison with the number that have fo-cused on women and their physiology, so two additional possibilities are vasopressin, the male equivalent of oxytocin, and testosterone, both of which, based on their established roles in child care, might reasonably contribute to the more rough-and-tumble interactions that many men have with their dogs.

Of course, we humans are never in complete thrall to our hor-mones: our vast prefrontal cortices allow us to make much more so-phisticated judgements about relationships than our cats or dogs are capable of. This raises a big question: if we really are capable of such a degree of rationality, why are so many of us such suckers for the charms of a demanding, messy fur baby? Perhaps our susceptibility to the charms of animals initially included far more creatures than the handful of species that we call 'pets', originating in the much more intimate relationship that our ancestors had with many kinds of wild animals and the natural world in general. It is here, way back in the history of our species, that we must seek the evolutionary origins of the hormonal responses that our pets trigger in us today.

The Evolution of Oxytocin

Oxytocin is an essential part of being a mammal: it is first and foremost a key regulator of parental care, enabling mothers to recognize and bond with their offspring. In chemical terms, it comprises nine amino acids, six of which are connected in a ring, leaving three to form a 'tail'. Its having evolved about 500 million years ago and remained almost the same ever since indicates its importance to mammalian life: it's identical in mouse and man, cat and dog. Without the surge of oxytocin that follows birth, mother mice fail to bond with their young (which they recognize largely by smell).

In some rodents, oxytocin has a second function: it cements the bond between male and female in species where both take care of the young. Again, a surge of oxytocin, this time immediately after mating, enables this (and again recognition depends largely on odour). In male rodents that form pair-bonds – prairie voles, for example – a post-mating surge of a very similar hormone, arginine vasopressin, performs the same function. Another rewarding neurohormone, dopamine, reinforces its effects.[28]

While the molecule itself hasn't changed, the way it works has, specifically in Old World primates (including ourselves). We recognize our partners by sight, not by smell, and the neocortex has all but replaced the olfactory area of the typical mammal (mouse, dog or cat): instead of following our noses, we look first and then think about what we've seen before leaping. As a result, although oxytocin is still important to our emotions and our behaviour, we've become somewhat emancipated from its effects. We can form social bonds in its absence (otherwise mothers would reject babies born by Caesarian section, for example). In order to achieve this new layer of social sophistication, primates evolved other ways of forming and sustaining friendships, chiefly mutual grooming and huddling together, and found new uses for other hormones.

CHAPTER 9

A Walk on the Wild Side . . .

PETS ARE ANIMALS first and foremost, even though they live alongside us in our homes. Perhaps they appeal in part – cats especially – because they bring a touch of wildness into our otherwise regimented and technology-dominated worlds. Long before domestication, our ancestors lived cheek by jowl with wild animals for hundreds of thousands of years and depended for their very survival on understanding them – which animals would poison or eat them, which would make the easiest meal. Our desire to interact with animals, far from being a modern affectation, is likely a legacy of our distant past, a set of responses developed way back in our evolutionary history, long before the first pet. Perhaps pets even fit within a broad set of evolved responses to the natural world that we all possess but rarely use nowadays.

Of course, we relate to pets very differently than to other animals. We may like observing gorillas in a zoo but don't want to take them home with us. We may delight in watching a soaring hawk, but few would desire to caress it. Perhaps pet owners are simply indulging in a kind of nostalgia for our ancestors' rural past, when cats and dogs shared our firesides as well as working for their living. If that were the case, however, pet keeping should have withered and died as we became irreversibly urbanized. This currently shows no signs of happening. Pets are at least as popular as they ever were, perhaps even more so. And their appeal runs deeper than any mere cultural craze or whim.

Certainly culture plays a role in our relationships with pets – and nowhere more clearly than in the animals we choose to take into our homes. One obvious example is the shift in dog breeding at the end of the nineteenth century, which split the species first into scores and then into hundreds of genetically separate breeds: had this not happened, the intense swings of fashion favouring different kinds of dog during the twentieth century would have been far more muted and might even have passed unnoticed. The very nature of pet keeping, of course, transformed radically over the course of millennia, as we moved from hunting and gathering, through settled agriculture and then industrialization, to the post-industrial societies of today. Along the road to their place by the fireside, dogs have been hunting partners, guards, livestock herders, turnspits and narcotics detectors – indeed new uses for their sensitive noses and powerful drive to win the approval of 'their' humans emerge year after year. Much more recently, cats have made the (awkward) transition from outdoor pest controller to apartment-dwelling companion. A kind of co-evolution, mainly between pets' genes and our cultures, has made pets what they are today – and therefore determined the options of the novice pet owner.[1]

So if culture is largely responsible for which pets we choose and how we relate to them, what explains our impulse to keep them in the first place? The answer depends on which conception of human nature we start from. The *tabula rasa* philosophy, which began with Aristotle and dominated twentieth-century social sciences, holds that the human mind is a blank slate on which anything can be written; this view suggests that if pets disappeared or were prohibited tomorrow, within a couple of generations the habit of pet keeping would disappear entirely, replaced by other fashionable pursuits. Unfortunately for its proponents, 21st-century neurobiology has debunked the *tabula rasa* idea (we inherit the shape and basic workings of our brains genetically; your brain is much more like your sibling's or your parents' than it is like that of the stranger you just passed in the street).[2]

An altogether different approach has emerged during this century, generally referred to as evolutionary psychology. Evolutionary psychologists attribute a biological basis to much of human behaviour. In their view, instead of being a blank slate, the brain consists of a number of specialized information processors, specifically those that helped our ancestors solve commonly recurring problems and thereby enabled them to survive and leave offspring – you and me, ultimately.

Evolutionary psychology starts from the premise that our brain evolved from that of our primate ancestors. It generally focuses on those categories of human behaviour that are more or less universal rather than those that clearly vary between groups and therefore plausibly stem from differences in culture. Thus, the field provides explanations for some aspects of religion, such as the near-universal belief in the supernatural, but not others, such as the predominance of specific religions in different parts of the world.

For behaviour to have persisted, natural selection must have come into play somewhere along the way. A major problem facing evolutionary psychology is pinning down the point at which a particular habit of ours might have evolved. It's extremely doubtful that the world Westerners live in today has shaped their evolution much, because it has encompassed too few generations. Since we believe that our species lived in small groups as hunter-gatherers for at least 10,000 generations, whereas only about 500 have elapsed since the dawn of agriculture, evolutionary psychologists usually turn to the former era when searching for adaptations. Hence we find the widespread but overly simplistic concept that humans still have Stone Age minds and appetites, accounting for our exaggerated hankering for sweet and fatty foods (previously scarce but nutritionally valuable and now the opposite) and for our enjoyment of canned laughter on television (before language, laughter may have signalled relaxed sociability).

Since isolated hunter-gatherer societies have survived in several parts of the world, we can test some of evolutionary psychology's predictions directly, bearing in mind that as very rare specimens of a

once-universal way of life, such societies may not typify how the average human lived even 10,000 years ago, let alone 100,000. Moreover, the notion of the 'typical' hunter-gatherer can be little more
than an abstraction, since hunter-gatherer societies varied dramatically from place to place. Survival in the northern subarctic (Inuit
and Yupik) entails very different ecological pressures from those encountered in Amazonia (Guajá and Yanomamo).

Comparisons with our closest living primate relatives offer one way
to try to understand the human mind, since we can observe their behaviour directly. The fossilized skulls of our hominid ancestors can also
be informative, even though we have only the sketchiest idea of how
they behaved. For example, the general consensus is that the main
evolutionary pressure behind the brain's massive expansion during the
apes' evolution was the increasing complexity of their social lives and
their need to understand others' behaviour and, for the more advanced
species, motivations. Scientists have compared our own capacity to
glean other people's intentions with those of various apes (for example, chimpanzees' abilities approximate to those of a four- to five-year-
old child) and then matched these with the volumes of the various
species' frontal lobes, where this talent largely appears to reside. As
the frontal lobe and other parts of the brain expanded, so did the
maximum number of individuals that a primate could recall and keep
track of (about 60 for chimpanzees, increasing to 150 in humans). Extrapolating from this data and incorporating the skull measurements
of extinct hominids, researchers have reconstructed the likely social
abilities of many of our ancestors and their close relatives.[3]

The greatest problem facing evolutionary psychologists is our very
uniqueness. Not only are our frontal lobes about five times as big as
those of chimpanzees – indicating a step change in evolution – but
they also appear to be organized a bit differently. Our mental capacities, especially for conscious thought, may have no close parallels
anywhere else in the animal kingdom. Perhaps this explains why
evolutionary psychology has enjoyed its most convincing successes
in elucidating human activities that are mostly biological in nature,
such as our sexual behaviour.[4]

Evolutionary psychology also explains our preferences for some kinds of natural surroundings in terms of their value to our ancestors. Some scientists have even gone so far as to claim that we have an innate love of the living world in general and of (non-harmful) animals in particular. First propounded by the Pulitzer Prize–winning biologist Edward O. Wilson in his book *Biophilia*, this impulse entails 'the urge to affiliate with other forms of life',[5] a somewhat romantic idea that some have employed to support moral arguments favouring nature conservation but that has received little direct support from psychology – except in one area: human preferences for particular kinds of landscape.

Our desire to go outdoors for recreational purposes plausibly reflects our hunter-gatherer past. Early research suggested that we might feel naturally at home in savannah-type landscapes, where our species probably evolved. Proponents of this hypothesis pointed to the recreational value of such open vistas found in the eighteenth-century English parkland designs of Capability Brown, which remain popular today. Central Park in New York is a classic example. However, subsequent studies suggest that the typical person prefers a balance between openness and concealment, which makes sense in evolutionary terms: our ancestors presumably valued openness because it allowed them to see their enemies coming, but they also needed hiding places for when those enemies appeared.[6]

Other kinds of landscapes suggest similar evolutionary biases. We enjoy wildflowers, which indicate the location of fertile soil. We like to be near rivers, lakes, and oceans, which provide water for drinking and/or fish for eating. Interestingly, our landscape preferences seem not to be entirely fixed but can vary with environmental conditions – for example, our partiality for landscapes containing water declines in wet weather. Of course, humans can come to enjoy new landscapes over time; for instance, an appreciation for underwater views is highly unlikely to be an evolved trait since our ancestors would have been unable to appreciate them.[7]

While we can adapt to value new kinds of landscapes, we do seem to remain drawn to nature, even though few of us experience it on a regular basis. More than 3 billion people worldwide live in towns and cities with limited or no access to 'nature'. Some might argue that this is an inevitable consequence of technological progress and that mankind will adapt to this change as we have to so many others. Maybe so – urbanization is a very recent event in evolutionary time. But we do not seem particularly inclined to embrace it. One look around a modern office will reveal that many people still include elements of nature in their everyday lives, be it a potted plant or a calendar showing natural scenes. This is not mere whimsy: indoor plants in offices or classrooms seem to reduce sick time, feelings of fatigue and stress, and physical symptoms such as dry skin, sore throats and coughing. Apartment-dwelling children who view green space from their windows have better cognitive abilities than those who stare out at masonry and concrete. In Toronto, one survey of the distribution of the city's trees forecasted that an increase of eleven trees per city block might increase residents' lifespans by sixteen months (on average). Even artwork depicting the outdoors naturalistically seems to confer mental health benefits: in one Swedish psychiatric hospital the patients occasionally attacked the paintings that decorated the wards, but records of the damaged pictures over a fifteen-year period revealed that they only defaced abstracts and had never touched a picture showing a natural scene.[8]

If just including a small slice of something natural into one's daily routine has benefits, then more direct involvement should offer even more. Belief in the healing powers of contact with wilderness is not new: nineteenth-century American philosopher-naturalists John Muir and Henry David Thoreau extolled the benefits of the outdoors and inspired the founding of the United States' national parks. Thoreau himself preferred open spaces to alcohol – 'who does not prefer to be intoxicated by the air he breathes?' – and, indeed, interacting with nature genuinely helps people in recovery from alcoholism and other addictions. Scientists have now documented many other benefits of natural surroundings, including enhanced psychological well-being (increased self-esteem; reduced feelings of anger, anxiety and frustration), greater ability to think clearly, and improvements in many physiological indicators of good health (reduced blood pressure and cortisol; faster healing from surgery). There are also social effects: when surrounded by nature, people tend to be more sociable and less inclined to violence. Exercise taken outdoors may enhance the health benefits of being active. Trekking, canoeing and cycling may be more beneficial than the equivalent amount of exercise performed in an indoor gym; even viewing pictures of natural scenes may improve the latter.[9]

Gardening, while not so strenuous as some other forms of outdoor exercise, may provide subtly different dividends. Community gardens seem especially beneficial in that they provide opportunities for social engagement in addition to exercise. In one study conducted in New York in the 1990s, two-thirds of community gardeners listed enhanced self-esteem as a primary reason for participating. Satisfaction from gardening may be a peculiarly human attribute, developed over thousands of years. Planting and tending provide delayed, not immediate, nutritional benefits and therefore require a degree of foresight and self-restraint not demanded by hunting or foraging. The beginnings of this ability may trace back to that point in the evolution of our species, some 30,000 years ago, when we began to manipulate the distribution of plants so as to provide more predictable food sources than we obtained from opportunistic hunting and gathering. Such 'natural' gardening was an important stepping-stone in the cultural evolution of our species. It led to the forager/horticulturalist lifestyle that characterizes many of today's small-scale societies and eventually, some 10,000 years ago, to the domestication of food plants.

Interactions with the natural world must have shaped our brains. Urbanization and industrialization have disconnected us from the wilderness that moulded our evolution, but many of us subconsciously crave contact with open spaces, as the popularity of outdoor pursuits and the value placed on urban green areas attest. Caring for some kinds of domestic animals inevitably brings their owners into contact with the outdoors – house pets not so much, of course, with the notable exception of dogs. Indeed, if some owners derive some health benefits from walking their dogs, then the most plausible mechanism is incidental exposure to green landscapes.

THE HEALTH BENEFITS obtained from engagement with the natural world seem robust (more robust than those claimed for owning a pet) and may tell us something about the evolution of our brains. However, we know little about the precise aspects of nature that our minds tune in to. We seem to draw the most benefit from more complex environments – for example, a city park that contains a mix of woodland, grassland and water, and therefore a wide range of plant species. The presence of animals seems less important, perhaps because few people notice them. One study, for example, linked the number of bird species to far fewer indicators of well-being than the number of types of plants – and butterflies seemed wholly unimportant. The smaller mammals tend to be nocturnal or to keep hidden, so are unlikely to have any impact in urban or suburban areas.[10]

Evolutionary psychologists have not paid a great deal of attention to our relationships with other members of the animal kingdom. However, those isolated phenomena that they have investigated help to explain some aspects of human behaviour that superficially make little sense in the world of today. One is our reaction to hazardous animals such as snakes and spiders.

The neurotoxic venoms of snakes and spiders affect almost all mammals, and so strategies for avoiding them should be evolutionarily ancient. In the West, snakes no longer pose much of a problem: in the United States in 2002, there were seventy-nine recorded deaths due to 'contact with venomous animals', compared to around 11,000 attributed to overdoses of narcotics and over 30,000 caused by shootings. If our minds were blank slates, we would fear guns much more than snakes or spiders.[11]

Most of us, however, are much more immediately afraid of snakes and spiders than we are of handguns (or syringes). Comparisons between the brains of the lemurs of Madagascar (where there are no snakes), New World monkeys (which have had intermittent exposure to venomous snakes over their evolution), and Old World monkeys and apes (which poisonous snakes have preyed on for millions of years) show that we have inherited visual systems that specialize in the rapid detection of snakes. Scientists have located these systems in the pulvinar neurons, which feed information from the retina and the superior colliculus (part of the midbrain specializing in visual input) directly to the fear-inducing areas of the amygdala. We do not make these connections automatically: young children and young monkeys alike find images of snakes fascinating and learn to fear them (see illustration). As they do not instinctively flee in terror from the sight of a snake, we can deduce that ophidiophobia, as it is officially termed, is acquired.

Fear of spiders (arachnophobia) has a similar evolutionary history. Our hominid ancestors probably evolved in southern Africa alongside highly poisonous widow spiders, many of which are black and rather small, tend to hide in crevices, and can be very difficult to detect. Correspondingly we appear to have evolved a specific and

Young monkeys find snakes both scary and fascinating.

very sensitive spider-outline detection mechanism. Our brains evidently invest heavily in keeping us alive by alerting us to poisonous animals.[12]

OF COURSE, OWNERS find their pets the very antithesis of noxious. Do we also possess evolved responses to other animals? Interactions with our own pets obviously bring us pleasure, but what about encounters with animals in the wild? Anecdotally, people describe contact with highly charismatic mammals as extremely memorable and even constituting a 'peak experience' – that is, among the most ecstatic, joyous, and happy moments in their lives. The popularity of safari holidays may testify to the psychological benefits of viewing large mammals. Many people take great pleasure in communing with animals in the wild. As one person wrote of a chance sighting of a wild dolphin, 'I was totally wrapped up with this experience. My heart was pumping. My heart was going so fast, but it was all enjoyable, it wasn't stress. It would be the epitome of joy in your life in a moment. But that moment lasted several minutes because I was walking swiftly. But those were incredible moments.'[13]

In some cases, people experience tremendous excitement even when the animals themselves aren't present. In one study, taken from accounts of wilderness trails in Namibia and the Kruger National Park in South Africa, participants rated finding signs left by large animals almost as highly as seeing the animals themselves, almost as if they were stalking their prey: 'We became so excited when three of us saw some rhino spoor, that we forgot about the time, and came back to the camp late.' Such reactions may harken back to the adrenaline rush our ancestors experienced during the hunt for prey animals.[14]

The number of people who visit zoos each year – around 160 million in the United States, over 20 million in the United Kingdom, and around 10 per cent of the world's population in total – demonstrates the popularity of large mammals. The need to establish some kind of contact with wild animals seems to remain strong for many people and has served as profound inspiration for some. Zoologist Desmond Morris said of his first trip to the zoo, 'That visit did more for my later interest in animals than a hundred films or a thousand books. The animals were real and near.' The emotional impact of viewing captive animals has received little study, but those who choose to visit zoos seem universally to voice wonder and respect. Indeed, anticipation of these feelings may have motivated many to visit. People with pets at home report more caring feelings for zoo animals than those without pets, but it remains unclear whether experiences with their pets affect their attitudes towards animals in general or they have a positive view of animals in general that both motivates them to keep pets and engenders concern for the care of zoo animals. Thus a general biophilia may account for the popularity of zoos, but the psychological roots of the phenomenon, at least with respect to animals, still require investigation. Nevertheless, our fascination with animals points to the likelihood that our brains evolved to tune us into the ways of animals that might affect our survival – positively or negatively.[15]

EVOLUTIONARY PSYCHOLOGY ALSO hypothesizes that the brain is not simply a general-purpose computing machine but contains modules specialized to solve specific problems that our distant ancestors encountered. Having learned more about the brain's construction, we now know that these modules undoubtedly exist but are not physically located in just one part of the brain. Instead, as a direct consequence of the brain's development, they are scattered around, as linguist Steven Pinker has graphically suggested, 'not like a flank steak . . . on the supermarket cow display . . . [but] more like roadkill, sprawling messily over the bulges and crevasses of the brain'.[16] Given the importance to our distant ancestors of beasts of every kind, our brains ought to have modules that specifically process information about the animals around us. Some of these might relate to interactions that few of us experience nowadays – threats from large predators, for example – but our brains, largely shaped by our ancestral hunter-gatherer lifestyle, should still retain traces of them.

    The ways we think about animals hint at how these modules might be working. When distinguishing between types of animal, our brains

seem to construct universal, and therefore presumably evolved, clas-
sifications. Visitors to a zoo may learn the names of all the animals
they view, in addition to much else, but these zoological categories
rest on centuries of scientific research. What of cultures that have no
science as such but depend for their very lives on intimate knowledge
of animals? Here human universals may point to a common cognitive
basis for categorizing nature in general and animals in particular.

There's good reason to believe that our thinking about animals
evolved along a different path from the one we use to think about
people. Given that we are the only bipedal hairless primate on the
planet, recognizing another living being as a member of our own
species is not difficult (though this may not have been entirely true
50,000 years ago when Neanderthals coexisted alongside our European ancestors or when other recently discovered hominins, the
Denisovans, occupied Southeast Asia). Identifying that other person as friend, ally or foe, potential mating partner or rival, however,
would have been crucial and relied largely on facial recognition.

With regard to other mammals, whether prey or predator, our
ancestors didn't need information about specific individuals. But
they certainly did need to know its species so as to anticipate its
behaviour: Is it dangerous? Is it easy to catch? Can I stalk it, or will
ambush likely be more productive? Or is it stalking me?

Again, studies of small-scale forager/horticulturalist societies have
given the strongest indicators of the human mind's predisposition to
learn about animals. Urban children in the West anthropomorphize
animals remorselessly, it turns out, because they have little experience to fall back on. Rural children still enjoy stories about animals
that talk, but at the same time possess a robust 'folk biology' based
on their day-to-day accumulated knowledge of real animals. They are
fully aware of the difference between Beatrix Potter's Pigling Bland
and a real pig in a way that many urban children seem not to be.

Children in small-scale societies, even those as young as four
years old, conceive of animals as entirely separate from humans; boys
often learn this distinction, as well as the differences between species, fastest because, unlike with girls, their fathers encourage them

to encounter first-hand all those animals that will become important to them later in life. At the same time they acquire a detailed 'folk ecology' relevant to their environment – for example, they know what each kind of animal eats and even which animals spoil the fruits they themselves would like to eat. By the time they are adults, their knowledge of the birds and animals around them will likely rival that of Western experts with a wealth of written and online information at their disposal. For example, Itza' Maya Indians (from Guatemala) classified photographs of birds from the Chicago area equally well as bird experts from Chicago classified equally unfamiliar lowland Guatemalan birds; American undergraduates were hopeless at both.[17]

Such folk biologies, or, perhaps more specifically, 'folk taxonomies', seem to be organized along similar lines wherever they have arisen and therefore probably involve an evolved trait of the human mind. Some variations arise with regard to the local fauna present and the priorities of the particular set of people involved; for example, although the Itza' Mayas display knowledge of many types of birds and can even guess the habits of those found in Chicago, they know most about birds of prey and game birds, such as the wild turkey: species important in their local economy and mythology. Nevertheless the general principles are usually similar. First, the human mind intuitively understands that species are immutable: a puppy raised with young piglets always turns into a dog, not a pig. Second, all humans seem to classify animals (and plants) into a natural hierarchy – as distinct from other, more arbitrary types of classification (such as colour or size). As Charles Darwin wrote, 'From the most remote period in the history of the world, organic beings have been found to resemble each other in descending degrees, so that they can be classed into groups under groups. This classification is not arbitrary like the grouping of stars in constellations.'[18] Thus a lion is a type of cat, which is a type of mammal, which is an animal. A red deer is a type of deer, which is also a mammal but is not a cat. Such relationships, nowadays formalized by genetics and evolutionary biology, also emerge spontaneously in the minds of people who have direct everyday experience of particular animals and live in cultures in which knowledge of those animals is essential to survival.

Folk biology parts company with modern biological science in locating the 'essence' of each type of animal. Science has shown that DNA, arranged into genes, encodes an animal's nature, but societies without such knowledge often locate 'essence' in the blood and/or heart, since the latter sometimes continues to beat in a newly killed animal. Such superstitions can live on even in modern cultures: some recipients of transplanted hearts believe themselves infused with aspects of the donor's personality.[19]

We once thought that children learned the basics of biology from adults and their own experiences, but recent research indicates that an inborn sense of biology enables them to separate animals from other things that move. By age six, most children in the West know that animals have complicated 'insides' essential to life and need food and water to keep going. Even at three years old, children understand that animals have 'insides' even if they don't understand precisely what they're for; for example, they know that a pig has more in common with a cow than with a china piggy bank. Remarkably, the beginnings of such awareness seem to surface in infants too young to speak. Already by six months they have acquired a social psychology that enables them to distinguish between people who help and those who hinder (in the experiments, the 'helpers' and 'hinderers' were animated wooden shapes with eyes, probably the minimum necessary to trigger the infants' concept of 'life'). By eight months old they also seem to understand that genuine animals are solid throughout: they show surprise when a self-propelling toy that has just had a 'conversation' with a person turns out to be hollow. If the toy is furry, they don't expect it to be interactive, just capable of moving without assistance; again, the infants remain perfectly content if the toy is solid but are startled if it's hollow. Even at that young age, they appear to follow a rule of thumb that animals move on their own and are either furry or capable of interacting with a person (in this case, their mother). If they have either of these qualities, they must also have 'insides'. Thus, from a very early age, children have an inborn concept of 'animal' that overlaps with but is distinct from their developing sense of how people are supposed to interact with one another – their 'folk psychology'. How fast the detail of their folk biology accumulates must depend upon their actual experiences with animals and the information they get from other people, accounting for the much more detailed knowledge of animals possessed by four-year-old Itza' Mayas as compared to members of the same age group in the West.[20]

The existence of different brain modules for folk biology (for animals) and folk psychology (for people) probably maintains the separation between the two. Given that we are such an intensely social species, our brains unsurprisingly give over large regions to analysing the behaviour of our fellow humans. The scope of some of these regions must extend to animals; otherwise we would not anthropomorphize them so readily.

Current research is beginning to pinpoint small areas within our brains that seem to operate as general-purpose animal detectors, quite distinct from those that respond to people or to dangerous creatures such as spiders and snakes. One study relied on patients with epilepsy who had not responded to treatment with drugs. Electrodes permanently implanted in their brains monitored the site of any subsequent seizures. Three brain regions – the hippocampus, entorhinal cortex and left amygdala – showed no selectivity towards pictures of animals, responding equally to images of people (mostly photographs of actors, sports stars and politicians). By contrast, in the right amygdala substantial numbers of neurons fired in response only to pictures of animals. The amygdalae are involved in processing information that relates to fear and threats, but this cannot account for the wide diversity of animal images that many neurons responded to since these included both dangerous (tiger, spider) and 'cute' animals (monkey, mouse) and even two cartoon 'animals' – Shrek and Yoda from *Star Wars*. In general, the study found no link between the number of neurons firing in response to a particular animal and how much the patients reported liking that animal or how exciting they found it. Furthermore, these neurons responded faster to animal stimuli than to others, suggesting that they were on some kind of neural 'fast track' for picking out animals from all the other visual information that streams into our brains every waking second. Intriguingly, some of the neurons seemed tuned to particular kinds of animal: one, for example, responded to rodents (mouse, rabbit, hare) but hardly at all to larger mammals (tiger, rhino). More animal-specific neurons in the brain await discovery in regions other than the amygdala: people whose amygdalae have been destroyed by Urbach-Wiethe disease show no deficit in their ability to pick out animals in complex scenes.[21]

Once our brains have deduced that an animal is approaching, they go on to gather as much as information as they can about the animal in question. Although still far from complete, the science has identified the anteromedial temporal cortex and the perirhinal cortex as two key areas crucial to distinguishing between different animals that may engender different responses: getting out of the way of a lion may not require the same tactics as avoiding a leopard.[22]

Selective detection of animals is unsurprising from an evolutionary point of view. Not only would our hunter-gatherer ancestors have had far greater exposure to animals than most of us do today, but their very survival would have depended on seeing animals first, before the animals saw them. Some would have presented danger by virtue of their size, possession of venom, horns or claws, aggressiveness or dietary preferences – at the time when the hominids were evolving, more species of big cat existed in Africa than do today, and fossil evidence indicates that our ancestors occasionally became lunch for a leopard, sabretooth or hyena. Others would have been potential prey, but the most valuable of these, because the largest, would also have entailed some risk for their hunters. Wild animals move unpredictably, more unpredictably even than humans, and therefore demand constant re-evaluation by the senses, of which vision would have provided the most detail. Nowadays, we might be much better off trading our animal-detection system in for a vehicle-detection system (a car causes one in every four accidental deaths in the United States). If our brains were not hardwired to a certain extent, we'd expect such an adaptation in today's adults, bombarded as we are with TV images of the damage that cars can do to fragile human bodies. Instead, our visual systems remain highly tuned to animal movements, even though most of us see far more cars than animals day to day. We are stuck with the brains that our ancestors evolved in a world in which wild animals played a huge part.[23]

WHEN DID THESE abilities first emerge? Humans have evolved 'modules' for interpreting the behaviour of animals distinct from those used to analyse the behaviour of our fellow humans, but is this true of other primates? Chimpanzees and other great apes not only lead sophisticated social lives but also appear capable of learning a great deal about the animals around them (other monkeys, in particular). Nonetheless, our abilities to reason about animals greatly surpass those of our primate cousins. At some point in our evolution our brains evolved specialized mechanisms that, beyond simply detecting animals, allowed us to analyse and predict their behaviour.

Intelligence comes in several flavours, in addition to the generalized learning abilities that all mammals share. One is social intelligence; technical ability is another; folk biology, a third. All may have originated in the evolution of our primate ancestors, with social intelligence perhaps emerging first. Chimpanzees have complex social lives but only a rudimentary tool-making ability, which some even use to exploit their knowledge of animals. For example, the chimpanzees of Gombe fish for termites using carefully selected twigs or grass stems, which they bite down to the right length and degree of stiffness. They then insert them into the ventilation holes in the termites' nest. The termites attack the intruding item, and the chimpanzee then pulls the twig – covered with clinging insects – from the hole and devours its meal. Despite this clever behaviour, chimpanzees don't seem to have a distinct technical intelligence. Their tool-making cultures are local – for example, many use stones to crack open nuts, but the precise manner in which they do so is often peculiar to their group – and they seem rather uninterested in teaching one another the tricks they have come up with, instead using their considerable general – blank-slate – intelligence to figure them out for themselves. Perhaps more tellingly, although they use weapons when fighting one another, unlike us they do not use physical objects as indicators of status or achievement – they have no badges of office or totem poles. Following their well-documented 'wars', chimps do not demonstrate their victories by parading the skulls of their victims.

Specialist technical intelligence is a hallmark of our own lineage. Our hominid ancestors began making tools from stone about 2.5 million years ago and developed far more sophisticated stone-based technologies than those employed by chimpanzees today. However, no further advances emerged for at least another million years, as if our forebears' brains somehow could not connect their technical intelligence to their ability to understand animals and their fellow hominids.

Indeed the Neanderthals, cold-adapted hominids who lived in Europe from about 700,000 years ago until their extinction about 35,000 years ago, may ultimately have perished because the architecture of their brains had not evolved sufficiently to connect their technical abilities with their knowledge of animals. In other respects they were rather like us: they had language, (eventually) mastered the control of fire, and (sometimes) buried their dead. They must have been sufficiently compatible with our more direct ancestors, *Homo sapiens*, who encountered Neanderthals when they emerged from Africa, to interbreed (humans of Eurasian descent boast 2 per cent Neanderthal genes). Yet they did not innovate to the extent that our *sapiens* ancestors did – they crafted no tools from antler or bone and little or no art. Moreover, while they must have had some knowledge of the animals they hunted, their hunting methods were relatively crude and largely confrontational, leading to many injuries and a correspondingly short lifespan. All this evidence suggests that, as with their ancestors from 2 million years before, their folk biology and technical intelligence remained unconnected.

The brains of our more direct ancestors may have changed to allow symbolic representations of animals around 150,000 years ago, soon after they left Africa and began to colonize the Fertile Crescent. Burials of early modern humans in caves in Mount Carmel (Israel) dating from around 100,000 years ago are probably the earliest to include wild animals – the head and antlers of a deer placed on top of a child, a man holding the skull of a wild boar. These may have been offerings to a deity (other, more nutritious body parts would presumably have been selected to provide food for the deceased person for his journey to the afterworld) or may have had some totemic significance lost on our modern minds. Whatever their precise significance, their inclusion in these graves indicates a shift in our ancestors' perception of animals to view them as more humanlike. Perhaps this marks the point when anthropomorphism first became mentally possible.

THE ABILITY TO out-think animals must have had profound consequences for our ability to obtain food, and not only through hunting. Small-scale societies notably have complex and highly developed relationships with the wild animals around them, perhaps dating back as far as 100,000 years. Symbiosis with an animal – specifically, a bird – once served even our sweet tooth, today responsible for a worldwide epidemic of obesity and diabetes.

We have always valued honey from wild bees highly. Nowadays, when honey comes in jars from farmed bees, and so many other sources of sugar are readily available, we can easily forget how much joy our ancestors must have felt when they managed locate and raid the honey from a bee colony. They did not always achieve this on their own: on many occasions a kind of woodpecker, the greater honeyguide, would have shared in the bounty.

Throughout the African savannahs, for many millennia, honeyguides and humans have worked together for their mutual benefit. Honeyguides don't care about honey. Like their woodpecker cousins, they eat the bees' grubs and any insects parasitizing the hive, such as waxworms. Unlike woodpeckers, they can also digest beeswax. The humans, however, are very interested in the honey, making this a win–win partnership.

We can't say where or how this relationship started or how widespread it became, though a few nomadic tribes in Kenya, including the Boran, were taking advantage of it up until the end of the twentieth century. Honeyguides go looking for bees' nests soon after dawn, when it's still too cold for the bees to be active. Later in the day, the bird goes searching for a person. Alternatively, when a Boran wants to find a hive, he locates a honeyguide by making a loud whistling sound, using a specially modified snail shell, a hollowed-out palm nut or, if neither is to hand, simply blowing into his clasped fists. Once the two have got together, the bird starts making brief flights in the direction of the nest, emitting a special double-noted call. If the Boran follow, the bird flies another short distance in the right direction and waits, still calling.

The human followers can judge from the bird's behaviour how far they will have to travel before reaching the nest. The further away the hive, the greater the distance between the bird's stops (perhaps it must check that it's heading in the right direction by flying on until it's sure and then doubles back to the rendezvous). The closer the bird gets to the nest, the shorter the distance between stops and the closer to the ground it perches. Once it arrives at the nest, the bird's call becomes softer, with longer gaps between the notes. The Boran then smoke out the bees and collect the honey, leaving the nest and its other contents for the honeyguide (see nearby illustration).[24]

This is a remarkably well-integrated relationship, although it is unclear to what extent long association with mankind has genetically altered honeyguides: perhaps every honeyguide must learn from scratch how to lead people. Certainly the birds do not always need people to obtain a meal: they also raid abandoned bees' nests (for their wax) or those already opened up by honey badgers. Honeyguides have reportedly guided honey badgers and even baboons; biologists have never confirmed this behaviour, however, so the ability to guide may not have evolved until the first modern humans appeared in Africa. Moreover, greater honeyguides show no sign of dying out in the many parts of Africa where shop-bought sugar has taken the place of wild honey, so they cannot depend wholly on their partnership with mankind.

An African honeyguide.

THE ARCHITECTURE OF our ancestors' brains continued to change over the following 50,000 years, judging not only from the shape of their skulls but from the increasing complexity of their behaviour, including their ability to anthropomorphize. Their hunting methods improved in sophistication and flexibility, becoming more effective than those of the Neanderthals and possibly contributing to the latter's eventual disappearance. Although rapacious hunting may be only partly to blame, the arrival of modern humans in Australia (50,000 years ago), northern Eurasia (40,000), and the Americas (16,000) coincided with the mass extinctions of many giant mammals, including five species of horses and mammoths in North America, mammoths and hippopotamuses in Eurasia, and marsupial 'rhinos' and giant wombats in Australia. These and many other species succumbed to the hitherto unprecedented abilities of modern humans to devise both specialized weapons (technical intelligence) and strategies (folk biology acting in collusion with folk psychology) to target the most profitable prey in the new lands they had discovered. Then, as the easiest prey disappeared, they presumably adapted again, focusing on less profitable smaller animals, until they reached some kind of ecological equilibrium, which then lasted until the invention of agriculture.

By about 30,000 years ago, the 'animal' and 'person' parts of our brains were evidently well enough connected for our ancestors to produce the earliest artistic depictions of animals. They painted mainly deer, bison and horses (but very few humans) on the walls of their caves; they carved representations of animals out of pieces of bone, horn and stone. Sometimes they went to great trouble, building scaffolding to reach the ceilings of large caves and working deep underground by the light of flickering torches. Scholars have long debated the purpose of such art. Was it part of a magic ritual, perhaps to secure success in hunting or to ensure the return of migratory herds the following year? Were the pictures illustrations for stories embedded in their cultural traditions and transmitted from one generation to the next? Were they instructional, used during the long winter nights to educate children in the habits of the animals they would hunt when old enough? All of these explanations may be correct, applying at different times and places. Some small-scale societies, notably the Aboriginals of Australia, still make cave paintings, and each image may communicate different meanings depending on the viewer. A picture of a snake beside a lake may indicate the risk of a poisonous bite in a particular place, or it may also represent the Wagyl, a snake-like creature from the Dreamtime (a zoomorphic mythology) that the Noongar believe formed many rivers and mountains in western Australia.

Anthropomorphism, a key to pet keeping, requires mentally mixing and matching the attributes of humans and animals. Our earliest evidence of this capacity comes from depictions of imaginary beings that are part human, part animal. One of the most famous, 'The Sorcerer', appears to represent a shaman. Found in a cavern at Trois Frères in France, it is thought to be about 15,000 years old. Interpretations vary, but it seems to include elements of several animals, possibly a reindeer and a horse, but with human hands and feet and the genitalia of a big cat. One of the earliest artworks ever unearthed encapsulates the blurring of human and animal natures: dating from over 30,000 years ago, an ivory statuette discovered in southern Germany depicts a half-man, half-lion figure (see illustration). Although, technically speaking, an example of zoomorphism (the depiction of mythical beings that are part human, part animal), the piece surely indicates that the artist also perceived some human attributes in the lion.

Such hybrid figures suggest that anthropomorphism was as much a human characteristic 30,000 years ago as it is today. However, whereas nowadays some regard the projection of human thoughts

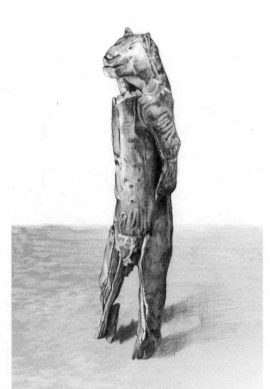

The lion/man ivory statuette from southern
Germany, some 30,000 years old.

and emotions onto animals as sentimental and frivolous, in the Paleolithic it would have served the serious purpose of predicting what animals would do in a given situation. This ability might have been helpful in outwitting large predators (which of those trees would a leopard choose to hide in?) but was probably even more useful in predicting how herds of prey animals (deer, bison, horses) might react when they noticed humans approaching with intent. It would also have assisted in designing effective hunting strategies, such as a coordinated ambush, that would cut one individual animal off from the herd to be killed. Although anthropomorphism is intrinsically only an approximation – no animal thinks in quite the same way as a person does – the detailed folk biology of these hunting peoples should have enabled them to allow for such differences when making predictions with a degree of precision that today's city dwellers can scarcely imagine.

One theory proposes that this new set of mental abilities resulted from an opening up between previously separate modules within the brain, perhaps around 50,000 years ago. According to this conception, early hominids, including Neanderthals, possessed three specialized intelligences, in addition to general intelligence: social intelligence, of which language formed an important part; technical intelligence, enabling them to make stone tools; and natural history intelligence, where their folk biology resided. During the evolution of our own species, these previously separate modules began to communicate with one another, producing a cross-fertilization of ideas and concepts that engendered many of the traits we think of as defining human nature (see nearby illustration). Technical intelligence interacting with social intelligence enabled the manufacture and use of symbolic objects, from bead necklaces to (ultimately) money. Technical intelligence married with folk biology enabled the invention of specific hunting tools, such as the Inuits' seal-hunting harpoon, which consists of over thirty components made variously from stone, bone, ivory, wood and leather. Folk biology merged with social intelligence would have provided the capacity for anthropomorphism, crucial for early hunter-gatherer pet keeping based on capturing young animals in the wild. The same connections, working in the opposite direction, permitted zoomorphism. All three would have come into play in the final great leap forward that eventually led to modern pet keeping, the domestication of animals: dogs are useful (technical intelligence) animal (folk biology) companions (social intelligence). It can be no coincidence that the first attempts to domesticate dogs, in western Europe some 30,000 years ago, occurred at the same time as the creation of the first zoomorphic art.[25]

THE RELATIONSHIPS BETWEEN early humans and wild animals, however harmonious and integrated, mainly reflect mankind's ability to learn how best to exploit the resources – including animals – around us. That information then gets transmitted from one person to another and, even without written language, from one generation to the next as an important component of a local folk biology. These relationships also reflect our unique capacity to 'think like an animal'.

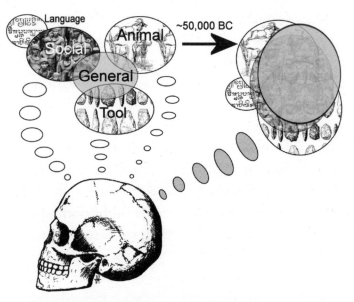

A facility for connecting the parts of the brain concerned with tools, animals, and humans may be a distinguishing characteristic of modern humans.

Even though we may only ever have an approximation of what is going on in the animal's head, anthropomorphism seems to work well enough to have enabled our hunter-gatherer forebears to exploit many different wild animals in diverse ways literally unimaginable to their hominid predecessors.

The human race's notably flexible behaviour has enabled us to adapt our habits to a wide range of habitats, from the equator to the Arctic, and thereby to colonize the whole planet. During early migrations we encountered countless thousands of kinds of animal and used our general intelligence and sensitivity to behaviour to exploit whatever we found. Some we rendered extinct; others found a way to coexist with us. But all remained wild until our ancestors caught the domestication bug and began to change the very nature of a few especially useful species. They cannot have always copied this habit one from another: several domestications of animals (pigs, cattle, sheep, goats) occurred in the Near East, but others came about quite independently in China (pigs again), Southeast Asia (chickens), South America (llama and alpaca), and North America (turkey). As the numbers of humans grew following the end of the last ice age, domestication of animals apparently became an inevitability. Our ancestors in several different parts of the world independently adapted their evolved ability to understand the ways of wild animals and finally brought a few of them under their control, thereby beginning the process that led to today's factory farming – and today's pets. Thus pet keeping is the child of our unique ability to understand the ways of animals. Thinking about animals is an intrinsic part of human nature but cannot wholly explain how our ancestors eventually brought some kinds of beasts under their much more direct control and thereby began to forge personal relationships with them.

CHAPTER 10

. . . and Back Again

MILLIONS OF YEARS AGO, our distant ancestors' brains evolved the ability to sense the comings and goings of wild animals, both those that might become a meal and those that might harm them. Today, most of us barely use these capabilities, surrounded as we are by flickering screens or speeding vehicles. The only animals that most urbanites encounter regularly, apart from birds overhead and the occasional visit to the zoo, are their pets. Descended from wild animals, pets trigger some of the neurons that our brains reserved as animal detectors, but they are not wild; they are very much a part of our society. Thus, while our brains may instinctively react to them as wild animals, due to our penchant for anthropomorphism we also conceive of them as part animal, part person. The proportion varies. A dog in the act of biting someone we view mostly as wild animal; a dog offering its belly for a rub we see as mostly person. Nevertheless, anthropomorphic thinking is never far away: after the event, we hold dogs that have bitten responsible for their 'crimes'.

The earliest domestication of animals, tens of thousands of years ago, required a huge conceptual leap: a new way of thinking about animals compatible with our dominion over them. Humans did not always see themselves as superior to, or able to achieve mastery over, animals. Indeed, some very recent small-scale societies hold very different attitudes towards the human–animal relationship. For example, animistic cultures such as the Native American Ojibwa traditionally conferred personhood equally upon humans, animals and

251

some inanimate objects, such as the wind – an essentially anthro-pomorphic worldview. They not only once interpreted personhood much more widely than we do but had no clear hierarchy for dif-ferent types of persons – they regarded rock-persons, bear-persons and human-persons as different but not unequal. As a consequence, they believed they had no authority to take any animal under their complete control – an attitude with echoes in the Buddhist/Shintoist traditions of Japan that to this day equate the use of a guide dog by a blind person with slavery. While small-scale societies keep animals as pets, these are not domesticated but individually obtained from the wild. Their pets rarely breed in captivity and therefore remain genet-ically unmodified. They are largely free to come and go as they please and thus do not violate an animist worldview.

So when did humans decide that they were superior to animals? Possibly around the same time that societies began to erect class bar-riers. Acceptance of the concept of domestication, dominion over animals, likely followed on from a shift in how people regarded one another well over 10,000 years ago. Most hunter-gatherer societies, it seems, were egalitarian, based on individual ties within each small community. For example, as recently as the late 1970s the Nayaka, a forest-dwelling hunter-gatherer tribe in southern India, lived in small settlements consisting of between three and five thatched huts. A single nuclear family occupied each hut but conducted much of everyday life in a shared outdoor space. Community members shared all resources, especially the spoils of hunting trips: this was a cultural norm not an edict handed down by a leader, of which there were none, at least according our own understanding of the term. A Nay-aka could only keep some prize to him- or herself by hiding it away or avoiding people entirely. Punishment, if caught, entailed ritual hu-miliation in the form of exaggerated requests to hand over whatever possessions he or she might have squirrelled away – the tribe's way of keeping the principle of universal sharing uppermost in everyone's minds.[1]

Not until the dawn of agriculture and the establishment of settle-ments did such ways of living begin to malfunction and our ancestors find that they could not avoid identifying certain individuals as leaders. Psychologically speaking, this change may also have allowed them to start perceiving animals as somehow inferior and therefore as potential possessions, rather than as their equals. The domestication of animals for their meat, milk, hides and eggs can only be the product of our modern, creative minds. No other member of the animal kingdom has ever domesticated another species; nor is there any evidence that our hominid ancestors did so.[2]

DOMESTICATION OF BOTH plants and animals began in the Fertile Crescent about 11,000 years ago, but the human mind seems to have been ready for a closer relationship with animals for some 20,000 years before domestication took hold (if the theory outlined in the previous chapter is correct). Dogs were undoubtedly the first animals domesticated. From time to time during those 20,000 years, isolated groups of hunter-gatherers tried to turn local wolves into something like a prototypical dog. At the time, unpredictable quantities of ice covered much of Europe, forcing people to migrate frequently. Dogs, which by that time had probably thrown off much of their wolfish behaviour and learned to attach themselves to individual people, may have been the only domestic animals mobile enough to keep up. By about 15,000 years ago, some had evidently succeeded in separating themselves forever from their lupine ancestors.[3]

Once the ice had retreated for the final time some 11,500 years ago, ushering in the Holocene era, our mental landscape was evidently well prepared for the domestication of animals. Of course, just because one person has a brainwave doesn't mean that everyone else immediately follows it. The concept of domesticating animals to serve our purposes, as an alternative to simply chasing after them, probably caught on very gradually, and migrations forced by the final phases of the last ice age may also have fatally interrupted early attempts at domestication. The first wave – goats, sheep (from mouflon), pigs and cattle (from aurochs) – began between 8,500 and 7,500 years ago, only a couple of thousand years after the domestication of wheat and barley. Moreover, these domestications occurred in the same part of the world, the Fertile Crescent, and so must logically be linked. Scholars still debate the precise connection and why they all occurred when they did, but the domestication of grains closely following the final retreat of the most recent glacial phase (the Younger Dryas) cannot be a coincidence.[4]

We can easily overlook the strong biological connection between domestic and wild animals. Today's domestic animals, from dogs to chickens, are so distinct from their wild ancestors that the uninitiated can find it hard to believe that they still belong to the same species (in the sense that the two can easily interbreed). Biologically speaking, dogs *are* wolves, and chickens *are* red jungle fowl. (Other domestic species, such as cattle and horses, have no living counterparts; their wild equivalents have gone extinct.)

Nowadays, all our domesticated animals seem isolated from nature. Their breeding, indeed their very existence, is totally under mankind's control. We have complete command over their genetics, having divided each species into many different breeds to fulfil the various purposes we have for them. In hindsight, it's easy to assume that each domestic animal must have been created at a single decisive moment, when one small group of people somewhere came up with the novel idea that rather than go out hunting, they could keep their favourite prey animal at home. If the captive animals thrived, they would spread through trade with neighbouring groups of people from their point of origin until they became ubiquitous.[5]

It is highly unlikely, however, that the early stages of domestication were as simple or straightforward as this presumption suggests. Nowadays, of course, most domesticated animals are completely isolated from their wild counterparts, although some live near enough to them to require physical prevention of interbreeding: all across Europe, domestic cats mate with the native wildcat, threatening the latter's genetic purity. Others, the mule being the best-known example, are incapable of breeding. Mules are a cross between a male donkey and a female horse, and because these two species have incompatible numbers of chromosomes, mules themselves are for the most part infertile. Other domestic animals are entirely isolated from the wild because their ancestors have gone extinct. All cattle descend from aurochs, the last known specimen of which died in 1627. The domestic horse's progenitor is extinct (its nearest living relative, Przewalski's horse, has a different number of chromosomes, indicating a long and permanent separation). Until recently, all golden hamsters kept as pets descended from a single female collected in Syria in 1930: this species was also thought to be extinct in the wild until discovery of a small number in southern Turkey about twenty years ago.

The isolation of domestic species from their wild forebears, however, in no way implies that the original break was quick or clean. For the past few centuries, all animal breeding has focused on choosing the right parents to produce the required kind of offspring – artificial selection. As the name implies, people control this process totally, with the animals having no say in which becomes a parent and which does not. In the early stages of domestication, however, our ancestors would have found it well-nigh impossible to exert much control over which animal mated with which. In those days, the gradual genetic separation of domestic from wild would have much more plausibly taken place through a process very similar to natural selection, Charles Darwin's 'survival of the fittest' (the alternative,

---

Sexual Selection

Biologists from Darwin onward have distinguished a third type of evolutionary process, sexual selection, which arises from the criteria females use to choose which male to mate with and vice versa. The male peacock's tail is perhaps the best-known example: its size is his advertisement to females of his ability to survive, despite carrying around what must be a considerable handicap except during the mating season. Modern domestic animals can't practise sexual selection because by and large they don't choose their own mates, but our selection for an unusual appearance is superficially similar – for example, we seem to prefer cats with white bibs and socks, even though these must betray them to their intended prey.

'sexual selection', is unlikely to apply directly to domestic animals see box). In this conception, some individuals – some wolves, for example – would have been better suited than others to living alongside man. Wolves capable of conducting themselves appropriately in human settlements might have bred with one another there, and any of their offspring that were less suited would either have left to (re)join their wild cousins or been killed. Either way, the choice of who mated with whom fell to the animals, and any genetic differences between the local wolves and the prototypical dogs would have come down to which individuals survived until they were old enough to breed.

Thus, the early stages of domestication were likely hit-or-miss affairs, as indicated by new genetic evidence from DNA profiling of both modern and ancient animals. The story of the domestic pig is a case in point. According to the conventional account, based on archaeology, several reproductively isolated populations of domestic pigs appeared in different parts of the world, within one or two millennia of the dawn of agriculture, and remained completely separate from their wild counterparts thereafter. But scientists now believe that the process was much more complicated and gradual. The latest research suggests that domestic pigs first appeared in Anatolia about 9,000 years ago and spread into Europe with nomadic swineherds. As they travelled, they interbred with local wild boar to such an extent that by 2,500 years ago virtually no trace of the original Anatolian DNA remained, having been supplanted by European wild boar DNA. Moreover, this DNA may have come from two different populations of wild boar: one that still exists in Europe and another, perhaps now extinct, from somewhere else, possibly another part of western Asia. Thus it seems that (Western) domestic pigs spent over 6,000 years in a kind of limbo, still freely interbreeding with their wild cousins.[6]

Continual interbreeding between domestic and wild, over thousands of years, raises awkward questions about how humans managed to control the animals they considered their property. Constant interbreeding with wild animals would have diluted any genetic changes that made some pigs more docile than others (for example), to the point where herding pigs would have been almost as impossible as herding wild boar. Some form of counterbalance must explain how pig keeping persisted throughout the six millennia of interbreeding.

Deliberate selection needn't only occur during the mating phase: it can also take place at any time after an animal is born. Pig DNA (from both European and Asian populations) suggests that continual genetic selection reduced the pigs' size to more manageable proportions and curbed their aggressiveness. Our ancestors probably could not have prevented a male wild boar from having his way with a domestic sow, but they could decide which piglets survived into adulthood. They could have culled any that showed signs of becoming too big or too pushy and permitted only the smallest and most sweet-natured to live long enough to breed.[7]

Mankind, it seems, played a remarkably haphazard but ultimately effective role in the early stages of domestication. Otherwise, we would not be able to trace the DNA of our domestic populations back to the dawn of agriculture. The process appears to have progressed via a combination of artificial selection – the culling of unsuitable animals before they posed a danger to their human hosts – and also some natural selection. Females tolerant not only of crowding but also of proximity to humans would have bred more successfully than their more stressed-out cousins, so genes for docility would have spread through the domestic populations.

CONVENTIONAL ACCOUNTS OF animal domestication usually focus on one species at a time, but the several near-simultaneous domestications that took place in the Fertile Crescent suggest that the very concept of domestication also spread. This cluster of domestications pre-dates written records, so transmission of the culture of animal husbandry could only have taken place by word of mouth. However, seeing or even just hearing rumours about animals under the control of nearby tribes might have encouraged settlements yet to acquire domesticated animals of their own to look to their local fauna for inspiration.

The most obvious prompt for domestication would have been right under their noses: the domestic dogs that were probably ubiquitous by 12,000 years ago. At that time, interbreeding with local wolves was probably a regular occurrence, and our ancestors might have extrapolated what they saw to other kinds of animals – if our dogs are a docile kind of wolf, then we might be able to tame some of those mouflons/aurochs/goats. Despite their antiquity, these modern humans had minds just like ours: there's no reason to suppose them incapable of making such a deduction. For example, other revolutionary features of the Neolithic 'package', like fired pottery and domesticated cereals, very likely spread outward, by word of mouth and example, from the site(s) of their invention.

The rapidly changing climate at the end of the last ice age initially hampered the domestication of animals, apart from the dog, but as humans living in the Fertile Crescent began to cultivate cereals and became more settled, it proceeded apace. As the first domesticated populations derived from wild animals in a particular location, interbreeding between the two would have been impossible to prevent and was therefore presumably frequent. Almost as quickly as we selected for the docility required to live in cramped conditions alongside us, matings with wild males would have diluted these characteristics. The resulting struggle to drive domestication forward may account for the seemingly hit-or-miss character of the creation of domestic animals.

WHAT UNDERLYING SOCIAL change drove the domestication of animals? It was probably not an overwhelming desire for meat, which was evidently still easy to come by through hunting: for the first five centuries or so after herding began, bones of wild animals recovered from village rubbish pits outnumbered those of domestic animals by five to one. Only later, after agriculture had become the dominant way of life and the human population had increased by perhaps as much as tenfold, did animal protein become harder to obtain.[8]

So, if not for meat, at least in the beginning, then why domesticate? For sheep, goats and cattle, one answer may be – for their milk. Aurochs and mouflon could be speared and roasted but would not have willingly shared milk intended for their own offspring. Recently scientists have distinguished traces of milk on fragments of Neolithic pottery: milk – sometimes fresh, sometimes cooked, possibly as a component of porridge – appears early in the Neolithic period, pre-dating settled agriculture. Bones of early domesticated cattle provide evidence of the deliberate killing of newborn calves, presumably to permit maximum exploitation of the mother's milk.

Early dairymen may have found milk less useful than we do today and may have fed it mainly to children and possibly other domestic animals. At that time, most adults would have had difficulty digesting more than small amounts of raw milk. Lactase, the enzyme that breaks down lactose, the sugar in milk, is highly active in infants, but most adults of African, East Asian, Melanesian or native South American descent are lactose intolerant, because after weaning, the amount of lactase people in these groups produce gradually declines. However, about three-quarters of adults of European or Near Eastern origin continue to make lactase for their entire lives, and the mutations that permitted this first began to spread among these populations about 10,000 years ago – just when the milking of domestic animals began. Right from the start, animal milk would have been a useful food for small children, allowing mothers to wean them early and free themselves up to perform other tasks. Had milk not been such an important commodity for our ancestors, this mutation would not have spread to the majority of the population.[9]

The selection of species for domestication may have also had a psychological dimension. The wild animals from which we derived domesticated animals already possessed totemic significance: aurochs figure prominently in Paleolithic cave paintings, and bas-reliefs of ferocious wild boar adorn the limestone megaliths of Göbeckli Tepe, at 11,500 years old the most ancient site for worship in Eurasia yet discovered (see nearby illustration). Swineherding plausibly began not so much to ensure a walking larder of pig meat but to display macho bravery, leading to social prestige: bacon might initially have been the bonus, not the impetus.[10]

Most of the domesticated species have become an integral part of our food chain, whatever the precise reasons for their selection.

Nowadays most of us maintain a firm distance, both physical and psychological, from the animals we eat. We make a hard-and-fast distinction between species that are pets and those that are food and feel uneasy when other cultures make different distinctions – hence the pressure on China and Korea to end the breeding of dogs for meat. Whereas few of us today are judged on our abilities with animals, only a couple of centuries ago most Westerners would have lived close to animals and been responsible for their upkeep. An aptitude with domestic animals was crucial to the development of our civilization – and of course remains important in many parts of the world. Those of us who so desire can still find an outlet for these skills in caring for our pets.

Wild boar feature prominently in carvings
on 11,500-year-old megaliths in Turkey.

OF OUR TWO favourite companion animals, the dog had already been domesticated by the time the Neolithic package of domestications took place. Cats, on the other hand, were only just beginning to inhabit our villages, first becoming domestic around 10,000 years ago – but probably not to the extent they are today. Until quite recently, scientists believed that cats first began to live with us several thousand years later – between 4,000 and 6,000 years ago, in ancient Egypt. Two fairly recent discoveries suggest otherwise. By sequencing cat DNA, researchers discovered that the separation between domestic and wild cats occurred much earlier, towards the beginning of the Neolithic period. The discovery of a cat buried, apparently deliberately, in a grave in Cyprus about 9,000 years ago supports the idea that cats became domestic far earlier than initially thought. The precise significance of the burial is difficult to deduce: we have uncovered no others, even though dog burials were commonplace by this time; furthermore the cat was buried close to, but not within reach of, its owner (if indeed that was the relationship between them).[11]

The most significant aspect of this discovery is its location: Cyprus had no indigenous wildcats, being too far from the mainland, and so the buried cat or its recent ancestor must have arrived there, in a boat, with people. There is a remote possibility that this cat left port as a stowaway and was adopted by someone on the ship, but a more plausible, if less romantic, explanation would be that cats, while not yet pets, were already serving as pest controllers on the mainland. We can be reasonably sure that the first settlers on Cyprus, around 11,500 years ago, accidentally introduced the house mouse to the island and may therefore have deliberately taken cats there a while later to help keep the mice under control. If true, this implies that the use of cats as in this capacity was already a well-established tradition around the Fertile Crescent and especially in Mediterranean ports. Thus the domestication of cats seems to have taken place in two phases: for the first five millennia or so they were simply pest controllers and possibly kept at arm's length; their transformation into pets probably did not occur until the second half of their association with mankind, around 5,000 years ago.

The history of the domestic dog is even more obscure, mainly because it is so ancient. At present, the DNA evidence is contradictory: some authorities place the site of first domestication in China, others in Europe, while still others maintain that DNA alone can never pinpoint either the time or the location with any accuracy. We can be sure that the dog was first domesticated thousands of years before the Neolithic revolution, and therefore by hunter-gatherers, not pastoralists, because the earliest ritual burials of dogs date from around 13,000 years ago.[12]

We can only speculate as to precisely why and how the unlikely relationship between mankind and its predatory rival began, since we have no archaeological evidence to speak of. My own preferred account is that a few wolves – of a type unusual for the time and now long extinct – began to scavenge around human settlements and that the offspring of the most docile of these became pets. Initially these adoptions may have been temporary, modern hunter-gatherer style, but at some point became more permanent and sufficiently durable to allow adult females to breed within the settlement and so start the process of domestication.

As with other animals, the process of domestication for dogs was long and complex. Several lines of evidence show that interbreeding with wild wolves continued for many millennia; for example, when the first humans crossed over into North America, the dogs they took with them hybridized with the local wolves, transferring a gene that renders their coats black, a trait that persists in roughly half the wolves in the United States today. Any genetic material transferred from American wolves into the domestic dog population evidently got selected out, as no trace of it remains even in ancient breeds such as the Mexican hairless.[13]

Whatever the details, much of the early history of the domestic dog seems to have been somewhat chaotic, with humans only exerting partial control over their reproduction. As with the continued interbreeding of pigs and wild boar, tame proto-dogs most likely became genetically distinct from the wild wolf through the selective culling of pups. Movement of some dogs – perhaps through trade – outside the wolves' normal range would have aided genetic separation. Overall, the irreversible separation of the dog from its ancestor would have been a gradual and erratic process, with some populations of domestic dogs lost to hybridization with local wolves and others abandoned as their owners moved north or south in response to the rapidly changing climate of the Upper Paleolithic.

Why our ancestors found these proto-dogs worth hanging on to remains obscure. They would not have been much use on hunting expeditions until they had become sufficiently trainable – which modern wolves, at least, are not – and the humans to whom they belonged had worked out adequate methods for harnessing their abilities. They may have initially functioned as guards, refuse disposals, and possibly beasts of burden: at one time Native Americans used dogs to pull travois as they moved from campsite to campsite. The nomadic Paleolithic peoples who domesticated the first dogs in western Eurasia might have benefitted from this type of assistance. Later on, certainly by about 12,000 years ago, dogs had become sufficiently well domesticated to assist with hunting.

This new partnership between man and dog may have had devastating consequences for other animals. Hunting dogs have been implicated in the extinctions of some megafauna that occurred after the colonization of new land masses by humans, though not in all: for example, the wave of extinctions in Australia began with the arrival of the first humans, about 40,000 years ago, and ended within 5,000 years, a full 30,000 years before the arrival on that continent of the first dogs. Stone-tipped spears were apparently sufficiently lethal to cause extinctions without canine assistance. However, it can be no coincidence that those occurring in North America some 13,000 years ago and in South America about 500 years later closely followed the first migrations of humans – and their dogs – to those continents (from Asia).

Like humans, cats and dogs needed to evolve psychologically for domestication to take root. Their resultant capacity to live among

The mammoth became extinct in North America within about
1,000 years of the arrival of humans and their hunting dogs.

and interact with us sets them apart from animals domesticated for
use as food, which need only be docile, provided they are also productive. Unlike their wild counterparts – and other domestic animals –
both dog and cat can take on dual identities. They retain much of
their wild behaviour: domestic cats can still become effective predators, as the birds nesting on a multitude of oceanic islands can attest,
and dogs, perhaps most obviously in those breeds used to herd sheep,
instinctively recall the very same pack-hunting manœuvres evolved
by their wolf ancestors. Both cats and dogs build social relationships
with members of their own species: while groups of dogs do not display quite the social sophistication of the wolf pack, cats cooperate
more with one another than their wild forebears did.

While cats and dogs retain certain aspects of their wild ancestors'
behaviour, both have an innate capacity not only to learn how to
interact with people but also to form permanent relationships with

particular individuals. Cats arguably remain territorial at heart and can only express affection for humans in places they know they can call their own. For dogs, however, relationships with individual people are essential for peace of mind. Neither kittens nor puppies are born with the psychological wherewithal to make these attachments: rather, they rapidly learn how to interact with humans during a sensitive period in their lives, when their brains are growing exceptionally fast (for cats this is between the fourth and eighth week of life; the equivalent period is rather later for dogs, between six and about fourteen weeks). Without the appropriate experiences of people in these brief windows, cats and dogs alike grow up to become permanently fearful of humans, or 'feral'.

This socialization is unique to these two species and is why we can leave them to roam unattended around our homes. Other domestic pets, such as rabbits, may appear to behave affectionately with their owners, but science has yet to show that they are capable of taking on a dual identity as cats and dogs do. Rather they, other small pets such as hamsters and gerbils, and outdoor 'pets' like horses become tame and habituated to the presence of people. Other domestic animals, such as sheep, can become tame to a greater or lesser extent, but there's no indication that they include humans in their existing social structures. We do not become 'part of the herd' as far as they are concerned.

Although we may talk of 'domestication' as if it involved the same process for all our household animals, it did not. Some animals we domesticated for food, and their journey into the modern world differed from that of animals, like dogs, cats and horses, domesticated for some job they could do for us. We turn only the latter into companions. In a sense, humanity has come full circle: hunter-gatherer societies indulged their need to keep pets by taking young animals from the wild, then domesticated certain animals following the agricultural revolution, and we now breed pets from our former four-legged helpmates. The species may be different, but our desire to make friends with animals has stayed the same.

DOMESTICATION OF ANIMALS changed how humans lived and eventually paved the way for both industrialization and our domination of the Earth's ecology. Did it also alter the structure of our brains? The occurrence of such a change over so few generations might seem implausible, but we know that the brains of cats and dogs must have altered radically over the same period; otherwise they could not have become socialized to us. At least one strong piece of evidence, lactose tolerance, suggests that domestication affected the physiology of our guts, so it's not out of the question that our brains also changed. However, that would not explain what induced us to bring animals under our control in the first place.

Pet keeping, which had probably been going on for tens of millennia, was plausibly the precursor to – and therefore is the key to understanding – the domestication of animals more generally. Looked at dispassionately, keeping pets is a rather unproductive habit, certainly by comparison with dairy farming. Its motivations are obscure but must have some evolutionary significance; otherwise it would not be so widespread. If pet keeping was the necessary precursor to domestication, then our brains were quite possibly already pre-adapted to nurture animals long before the domestications that were the most revolutionary expression of this particular quirk of human nature.

CHAPTER 11

So, Why Pets?

Pᴇᴛ ᴋᴇᴇᴘɪɴɢ ɪꜱ so widespread and so timeless that it must surely have roots in the evolution of the human mind. Of course, at any given time and place, economics, technology and culture largely determine the type of pets that people choose to keep. None of these, however, can plausibly account for the habit itself, which must therefore stem from who we are, from human nature itself.

Pets might plausibly fit into our evolutionary story in several ways. I propose that the pet-keeping habits of many hunter-gatherer societies were an essential precursor not only to pet keeping as we practise it today but to the domestication of animals. As an almost accidental spin-off of the pleasure they derived from raising the young of wild animals, a few of our ancestors began to exploit captive animals for their meat, milk and hides, facilitating domestication and kick-starting the agricultural economy that still dominates much of the globe. If this is true, our desire to keep pets is the present-day expression of a trait that served our forebears well.

There are, of course, other potential explanations for modern pet keeping. Perhaps pets are nothing more than hangers-on, an error of judgement on our part, or a misdirection of our desire to take care of one another. Perhaps we are somehow (and this also would require explanation) simply too undiscriminating to confine our affection to our own species. Some people do indeed argue that pet keeping represents a grossly ill-judged misdirection of resources that we should properly reserve for humanity. Some maintain that the purchase of luxury items for pets is morally indefensible in the face of extreme poverty in many parts of the world. Others focus on the environmental impact of pets, claiming that just feeding two large dogs takes up more resources than the average Bangladeshi uses in a whole year. Cats may be less resource heavy (they require about 10 per cent as much as a large dog), but their predatory habits incur the wrath of conservationists and wildlife enthusiasts worldwide. *Cat Wars: The Devastating Consequences of a Cuddly Killer* could be a tabloid newspaper headline but is in fact the title of a recent book by the director of the Smithsonian Migratory Bird Center.[1] A pervasive belief even holds that excessive empathy with animals can cause people to turn away from showing concern for people. Critics widely levelled this accusation at the nascent humane movement in the nineteenth century – as if empathy were some kind of currency that could only stretch so far. Nowadays, the fashion for fur babies attracts equal derision, as in the following from the New Modern Man website: 'Dogs are now taking the place of human babies in baby carriages. There is no bigger sign that your civilization is facing some serious issues with motherhood than when this becomes a popular trend.'[2]

Critics of pet keeping can easily find examples from nature in which other species get suckered: we are not the only animals who raise others' young. Perhaps the best-known instance is the cuckoo, which lays its eggs one at a time in the nests of other types of birds. Evolution has designed cuckoos' eggs to match those of their hosts; hence they parasitize particular bird species, warblers especially. The female cuckoo waits until her chosen pair of warblers have started laying and then sneaks into their nest while they are elsewhere. Within a few seconds she will have deposited her own egg, after cunningly removing one of her hosts', so that the total number remains the same. Once it hatches, the cuckoo chick grows rapidly. To guarantee that it gets all the food that its foster parents bring to the nest, it rolls the hosts' eggs out one by one until it is the sole surviving occupant. Unlike the egg it hatched from, the cuckoo chick looks nothing like a baby warbler, but it tricks its hapless hosts into feeding it with a vociferous call that imitates a whole nestful of chicks and with the bright orange-red colour of its 'gape' – the skin inside its wide-open beak.[3]

Today's pet owners sometimes get compared – unfairly, in my opinion – to the warblers who seem so readily duped by their cuckoo parasites. Opponents of pet keeping sometimes blame the trend among singles and couples to keep pets while they delay having their own children – more plausibly explained by economic and societal pressures – on the pets themselves, as if the animals were cuckoos, hijacking and then manipulating parental instincts. Of course, no one is suggesting that pets actually kill off existing children the way that cuckoo chicks do; however, these critics do raise the possibility that the very availability of pets stops some people from starting families. Others go further, complaining that adding a pet to a household that includes children siphons away the parents' time and money, both of which would be better spent on their own flesh and blood. To account for this apparent 'waste', they propose that some people are sufficiently susceptible to the attractions of pets that they fail to recognize this fundamental mistake, even though our big brains should give us an advantage over reed warblers.[4]

If pets are really just like cuckoo chicks, then it follows that owners must be as oblivious as reed warblers to their error. Unlike warblers, however, owners know perfectly well that their pet is not a human infant. Of course, humans do lots of illogical things, despite knowing that they shouldn't. Consider the current epidemic of diabetes and obesity, largely driven by our Stone Age preference for sweet tastes, despite a torrent of contrary advice from the medical profession. Addiction to heroin provides another example: cravings often trump common sense.

So if pets really are the ultimate con artists, then why have we not developed ways of calling their bluff? Unfortunately, evolution is a far from perfect process, and sometimes behaviour that actually harms the person (or animal) performing it can persist for long periods, even forever. Why have the cuckoo's hosts not learned to tell the difference between their own nestlings and the cuckoo? Some species may have – in Scandinavia, chaffinches and bramblings, although suitable hosts, are now rarely parasitized. For reed warblers, the risk of mistaking their own chicks for a cuckoo's and abandoning them unnecessarily may outweigh the costs of raising one cuckoo chick in their whole lifetime.

Another constraint is that evolution is 'blind' and only works on existing variations. As an intensely social species, we need to know what other people are thinking about us, and this is almost certainly the main reason why we evolved such big brains. At the same time, our standard mammalian senses – mainly sight and hearing – limit the amount of information we can gather, and so we can only guess others' thoughts through subtle and potentially misleading clues based on body language. How much simpler it would be if we possessed telepathy or some other form of extrasensory perception – once a familiar theme in science fiction.[5]

Of course, evolution will not have eliminated all traits that hurt us. Some are just too important to our survival, even though they occasionally present us with problems. Grief – whether for people or for pets – seems unnecessarily bad for us, leading as it can to months of impaired psychological and physiological health. However, under most other circumstances, a feeling of unease when we become separated from someone we know and trust, especially a close relative, spurs us to find that person; important enough today, this reaction was probably essential to our survival 50,000 years ago when the other members of our tribe provided our best, maybe only, protection. This system inevitably goes into overdrive for a while after the attachment figure dies. Somehow evolution has so far failed to connect our conscious ability to conceive of the finality of death with the much more primitive system that generates the distress of separation. It would be much neater if we could make up our minds to draw a line under grieving when someone has died, but grieving appears to have to take its own course.

The second major constraint on evolution is the speed – or rather lack of it – at which natural selection works. Variations in behaviour must pass through many generations before one that confers a particular advantage can significantly outperform the rest, and that can only happen if conditions – the selection pressure – don't keep altering. The pace of change in today's world is unprecedented; that's why evolutionary psychologists looking to explain some quirk of human behaviour usually start by imagining that it evolved during the hundreds of thousands of years when we and our hominid ancestors lived as hunter-gatherers in Africa and only, on not finding the explanation there, turn to later phases in our ever-accelerating evolution. Thus our craving for sweet tastes was once adaptive, when sources of sugar were hard to come by, and even though the trait is now not only maladaptive but unevenly distributed genetically – for example, Pacific Islanders are especially susceptible to diabetes – too few generations have elapsed since refined sugar became cheap for any significant advantage to accrue those who happen to prefer savory flavours over sweet. If (hypothetically) the twentieth-century boom in pet keeping were a form of parasitism, then we would not have had time to evolve defences against it: too few generations have passed since we began to keep dogs and cats purely for their company.

As proponents of the 'pets are parasites' theory would have it, our evolutionary predispositions dupe us into taking care of animals with no other function beyond providing (in their view, imaginary) companionship. Certain animals, and dogs and cats especially, first trigger our parental instincts and then (somehow) persuade us that they are not only irreplaceable but priceless: in a Gallup poll conducted in 2001, 44 per cent of US pet owners said that they wouldn't trade their beloved pet for any less than $1 million.[6] Globally speaking, both dogs and cats greatly outnumber their wild ancestors. Thus, by becoming pets, wolves and wildcats improved their chances of leaving offspring manyfold. They achieved this by tapping into our predilection for cuteness, both the *Kindchenschema* and 'whimsical' varieties, that evolved to ensure that we will take care of our own offspring. Again, according to the 'cuckoo' theory, our anthropomorphic instincts and our unwillingness to see the relationship otherwise fool us into believing that they love us 'unconditionally'.

If the detractors are right and pet keeping is really as silly and damaging as they claim, natural selection should have eliminated it tens of thousands of years ago, long before dogs and cats were domesticated. Since many neo-Paleolithic societies practise something rather like pet keeping, it appears to be an ancient trait – certainly old enough for natural selection to do away with it if it really were harmful. True, the species and breeds of animal involved differ considerably between societies and from one era to another, but the habit of adopting and caring for the young of other species seems as close to a human universal as most such habits are. Why, then, did the impulse to keep pets survive? Why, specifically, do we find baby animals as cute as human infants?

Cuteness certainly works. When it comes to human infants, it can trump our kin-recognition mechanisms – as indicated by the fact that fostering of young children is commonplace in small-scale societies. This habit benefits the biological mother, whose lifetime reproductive success increases if others foster some of her children. The costs to the fosterers, often grandmothers past reproductive age, are typically not too severe. However, all concerned know how closely related they are to the children they are caring for – so fostering contributes to reproductive success in a broad sense. And ultimately, cuteness alone doesn't win the day: some evidence suggests that women behave differently towards foster children, to the extent that foster children are more likely to fail to thrive when times get tough. It seems implausible that a human mother, so discriminating when it comes to human infants, could fail to recognize that a baby monkey or peccary is neither her own child nor the child of an unrelated member of their band.[7]

While cuteness may affect humans, it does not overwhelm their senses – particularly when it comes to animals. Whereas viewing pictures of cute puppies and kittens triggers some of the same brain regions that respond to human infants, other regions respond only to our own species. Likewise although there exists overlap in the brain's responses to children and to adult dogs, centring on affection, there are also differences. Owners' use of a form of baby talk when addressing their pets has generated the assumption that the interaction is triggering their 'baby pattern' reaction – but adults use varieties of baby talk widely with other adults, for example, lovers. Moreover,

mothers instinctively try to teach their babies the names of the objects around them; owners don't do this when talking to their pets.
In other words, they are aware of the difference between pets and
human infants, apparently without needing to stop and think about
it. Owners may describe pets as 'one of the family' to indicate the
depth of feeling that they induce, but no one is genuinely confused
about which family members are human and which are not.

Rather than glibly comparing children and pets, it's fairer to say
that dogs and cats have their own appeal, which overlaps with but
remains distinct from our feelings for our fellow humans. By comparison with their wild ancestors, dogs are certainly childlike in the sense
that they never quite grow up. Much of their behaviour evolved from
that of juvenile rather than adult wolves, and in this way they retain
some of the wolf cub's cuteness (the main exception being, of course,
that they mature sexually; otherwise they'd never be able to reproduce). Recent fashions in dog breeds also demonstrate our attraction
to the baby-faced types – pugs and Pekinese – as exemplified by Doug
the Pug overtaking Grumpy Cat as the most popular celebrity 'author' and animal star of the Internet.[8] Thus some, but by no means
all, dog breeds have undergone similar evolutionary changes to the
teddy bear, as a certain type of owner attempts to mould them into the
image of a cute toddler. Many dog breeds, however, are not in the
least babyish – Great Dane or saluki, anyone?

Cats' faces already projected the fur baby image long before
domestication, so they had less changing to do on the outside. On
the inside too, they may be little altered from their wild ancestors,
apart from the essential adjustment of learning how to socialize with
people. Even their apparent playfulness is something of an illusion,
since their 'toys' are, as far as the cat is concerned, more a prey substitute than the focus of a game: while an owner may think her cat is
playing with a toy mouse, in his head he's hunting. (With dogs such
play is generally a social occasion, as shown by the way they spontaneously present their owners with toys to get a game going.)[9]

The 'cuckoo in the nest' hypothesis is superficially plausible because the appeal of dogs and cats does have some features in common

with those that attract us to infants. However, it is intrinsically implausible because, unlike the cuckoo chick's hapless hosts, pet owners understand fully that their animal companions are not their children. Some may see their pets as child substitutes, but that is not the same thing as believing that they *are* children. But if pets are not parasites, what are they? Why do so many of us find them so appealing?

THE CUCKOO HYPOTHESIS fails to take into account the possibility that the traits that make dogs and cats appeal to us might have evolved not to exploit our good natures but to make these and other animals more useful to us. Nowadays both dogs and cats are primarily companions, but for most of the thousands of years that they have lived with us, they have performed tasks. Perhaps we find them appealing nowadays because our ancestors who did so fared better than neighbours who failed to appreciate them. In exploring this idea, we need to consider cats and dogs separately, because historically they have served very different functions, even if nowadays we treasure both for their companionship.

People who don't much like dogs sometimes point to their subservient natures as evidence of their inferiority. However, the very fact that dogs are so attentive to humans, and so rewarded by any attention they get from us, has made them trainable. The untrained dog is a liability today, and probably was historically, since it must be kept tied up or contained in some way to stop it from stealing food or harassing livestock and children. A dog not trusted enough to be let loose is useful only as an early-warning system. Uncontrolled dogs that do not fear people are a menace, not least because in many countries, even today, they transmit rabies: this may account for their negative portrayal in all three monotheistic religions, a viewpoint that persists in Islam to this day. However, a trained dog can be useful in many different ways, especially as an aid to hunting and herding or for protection. Indeed, people are coming up with new purposes for trained dogs even today, especially those that rely on the dog's nose, tens of thousands of times more sensitive than our own. For example, the US Department of Agriculture now uses Jack Russell terriers to detect brown tree snakes hidden in cargo being loaded onto planes in Guam (while poisonous, these snakes are not especially dangerous to either dogs or people; rather, they pose an invasive threat to wildlife).[10]

Mankind has valued cats for the opposite reason – for their independence. Having domesticated them primarily as vermin controllers, we have allowed – nay, encouraged – them to get on with their work in an entirely spontaneous way. Cats hunt best when we're not around to disturb their prey, and their agility and suppleness allow them to get into spaces that we cannot (but mice and rats can). However, to effectively control vermin, they have to at least tolerate the proximity of people. Wildcats of all species fear us too much to exploit the concentrations of prey around our dwellings. Thus, the apparent subjugation of an animal to mankind's demands has been, for the domestic cat, entirely voluntary and beneficial to both sides.

Cats and dogs have co-evolved with us over the past 10,000 to 15,000 years, albeit differently. Both have changed genetically to adapt to the opportunities – niches – that man has offered in ways that no other species has managed. We, in turn, have evolved culturally – especially in our breeding, husbandry, and training of dogs – so that we and they can coexist productively and fairly harmoniously alongside one another. Neither cats nor dogs have exploited us since they became domesticated; nor have we unilaterally exploited them as we have those animals domesticated as sources of food. Rather, we have evolved partnerships with them that have combined the abilities and talents of each for the benefit of all. They have prospered while their wild counterparts struggle to survive. We can reasonably assume, although we have no conclusive proof, that human cultures with cats to protect their food stores and well-trained dogs to guard their livestock have out-competed those with neither. Our relationship with both species, far from parasitic, has been much closer to symbiotic.

HUMANS AS WELL as cats and dogs have benefitted from a long history of association. But what about pet keeping specifically? Has our traditional, symbiotic working relationship with these animals, now separated from its functional roots, become parasitic in the biological sense? I'm not suggesting that dogs and cats are laughing behind their paws at our gullibility. (They're smart, but not that smart.) We bear the responsibility for this situation, since in the course of exploiting them, we enhanced those features of their appearance and behaviour that make them far more appealing than any wolf or wildcat. To refer to them now as parasites would be grossly unjust. However, one could still argue that pet keeping, divorced from other functions, is maladaptive in the here and now, when most Westerners keep dogs and cats purely for companionship.

Studies of relationships between brood parasites – like cuckoos – and their hosts show that hosts constantly try to find ways of discriminating between their own and the parasite's young. Given our huge brains, why are we seemingly incapable of reserving our affections exclusively for members of our own species? The answer may lie in our long and mutually beneficial relationships with both dog and cat. Both species have been useful to us, so perhaps a failure to understand their behaviour or resistance to their charms put people at a disadvantage. In that case there would have been less selection for, and possibly even some selection against, the evolution of a brain mechanism that prevented people from bonding with these animals.[11]

Since the 1970s, many studies of the social roles of dogs (and sometimes cats) have emphasized their potential benefits for our physical and mental well-being. Perhaps this is the necessary counter-argument to the 'cuckoo in the nest' hypothesis. As fewer and fewer cats earn their keep as hunters, and fewer and fewer dogs work for their living, have these animals neatly stepped into other roles that restore their usefulness to their owners and to society in general?

Nowadays, the media widely portray dogs (and sometimes cats) as the ideal antidote to the stresses and strains of everyday life, especially loneliness. Even if genuine, these effects are unlikely to have sustained the appeal of pets down the ages, since the problems they supposedly neutralize are, by definition, recent. The women and children of neo-Paleolithic societies have no need to counteract loneliness, since they rarely leave one another's sight; yet pets are often an important component of their lives. Thus the effects of pets on loneliness and stress, even if as great as their supporters claim, cannot explain, from an evolutionary perspective, why we find certain animals so appealing, as natural selection cannot foresee what kinds of conditions might prevail in the future.

WE MIGHT LOOK at pet keeping as no more than an evolutionary accident that's never been selected out, because, while we may indulge in it when we have surplus resources at our disposal, we can abandon the practice whenever it becomes a problem. I will go further, however, and propose that it's been positively selected *for*, because in the past it has afforded to those who happened to be good at it an advantage – or rather two advantages, each at a different stage of the evolution of our species.

It seems plausible to me that pet keeping served an evolutionary function in preparing girls for motherhood. Girls who learn to care for animals might draw on those skills later when looking after their own infants. No one has tested this hypothesis directly, but such a benefit might logically have contributed to the evolution of pet keeping: mistakes made when looking after a young animal are far less costly to an individual's lifetime reproductive success than those made when caring for an infant. This theory might go some way to explaining the fact that, on average, interest in pets is highest in prepubescent girls; perhaps less consistent, though not incompatible, is our continued affection for pets throughout our lifespans, among both men and women.[12]

What of the observation that pets make us more attractive to others? People have long used pets as status symbols, probably going back much further than the first toy dogs. Pet ownership could have signalled that a person had no need to toil – the equivalent of the ancient (and modern) Chinese habit of growing long fingernails. However, this may not have been its first or indeed most significant connotation. In hunter-gatherer societies, girls would have looked after baby animals not in isolation but in full view of the rest of the

In the Paleolithic era, adolescent girls who
showed an aptitude for caring for young animals
might have won preference as brides.

tribe. Those who observed them – especially potential fathers-in-law,
who likely had the most say in whom their sons married – may have
surmised that any caretaking skills they acquired would make them
better mothers when the time came for them to have children of
their own. Girls who were skilful and responsive keepers might have
acquired more value as potential brides than girls who were less attentive to their pets, to the point where those animals failed to thrive
or wandered off. Now that we in the West believe that individuals
should choose their own marriage partners, facility with animals
may have little influence, but during much of our brain's evolution,

parents likely made most nuptial decisions and perhaps used a facility with young animals as one criterion.[13]

One of the most consistent findings in the 'pets are good for you' literature supports the idea that a knack with animals has been important in mate selection: people get rated as more attractive when in the company of animals (under their control) than when they're on their own. This effect seems to apply equally to both sexes, as both subject and observer – whereas the partner-selection hypothesis predicts that it should be most powerful when men are selecting between women with and without pets. Indeed, the ability of a young man to obtain telephone numbers from young women when he was accompanied by a friendly black mongrel dog was among the most compelling results, as was the powerful effect of adding 'with a dog' to a description of a man on women's ratings of him as a potential partner. Thus women also seem to value men for their ability to care for animals. This trait, too, may also derive from early partner-selection practices: recent studies of small-scale societies suggest that we have previously underestimated the role of mothers and daughters in choosing potential sons-in-law/husbands.[14]

Partner selection in all its forms appears to have exerted a powerful influence on human evolution, especially in combination with cultural changes. In the scenario proposed here, parents would not have acted in isolation when choosing a bride who was good with animals for their son, or indeed vice versa: the other members of their group would have noticed and perhaps discussed their decision. In this way, such an initially one-off preference could have developed into a cultural norm through the widespread – and genetically based – human habits of imitating peers and conforming to social rules, thereby reinforcing selection of the trait 'good with animals'.[15]

NATURAL SELECTION COULD also have shaped a facility with animals. Domestic animals gave our ancestors an advantage during the Mesolithic and Neolithic periods, as they began to settle and claim the most productive land. Those with fertile soil and domestic animals would have thrived at the expense of those who persisted with hunting and gathering. But how did they retain control over their newly domesticated stock, given that interbreeding with wild forms would still have been commonplace?

The early history of animal domestication contains a paradox: the process, for all its evident advantages, seems to have been a highly hit-and-miss affair, often failing. And it took many centuries for even successful efforts to come to fruition. Societies that encountered domestication fully formed, however, seem to have adopted the practice immediately and enthusiastically. For example, the indigenous peoples of South America replaced their 'Aguara dogs' – semi-domesticated foxes – with dogs obtained from the Spaniards within just a few decades of their becoming available. Likewise, many neo-Paleolithic societies keep domesticated chickens and pigs based on stock originally supplied by outsiders. All this suggests that while acquiring the knowledge necessary to look after an already domesticated animal is easy, domesticating a new species is a long drawn-out and difficult process, which may require a very special kind of person for sustained progress.

Cultural factors would then, as now, have been extremely important. None of the ancient domestications could have been accomplished within a single lifetime, let alone a single generation. It would have been imperative for each generation of neo-pastoralists to pass to the next its accumulated knowledge of how to look after animals, along with the animals themselves. If this process broke down even once, the part-domesticated animals and thus all the genetic improvements up to that point would have been lost, and the whole process would have had to start again. Many such disruptions would have lain outside human control, especially in the unpredictable climatic conditions at the end of the last ice age. Either a sustained drought or a few years of cold weather, causing extreme hunger or forced migration to more hospitable habitats, could have interrupted incipient domestication, either precipitating the demise of the animals themselves or breaking the transmission of the detailed knowledge of how to care for them, or both.

Different groups' reactions to such a catastrophe would most likely have depended on how they regarded their animals. Those who viewed their semi-domesticated animals as mere commodities might have rashly killed them for food, unaware that they would be impossible to replace. If, however, some individuals within the group regarded their domestic animals with deep affection – more or less, as pets – then they might have preserved those particular animals as personal, rather than communal, property. The emotional bond between owner and 'pet' could have sufficed to spare the animal, preserving those genetic changes that already taken it part way down the road to domestication. This decision might have involved an element of foresight that the group could breed the remaining animals again when the situation improved, but even if it didn't, the net result would have been the same: preservation of the precious domestic genes.

THE HUMAN IMPULSE to keep pets may also help explain another awkward fact about most domestic animals: their DNA indicates an implausibly early separation from their wild ancestors. By definition, domestications can only take place in areas inhabited by those ancestors, and in the early stages, wild versions would have greatly outnumbered the domestic. Thus interbreeding between wild and domestic would initially have occurred on a regular basis – especially among wild males attracted to domestic females in season – thereby diluting and potentially undoing all the genetic changes required to make the species sufficiently manageable.

We see little of this interbreeding nowadays because the wild ancestors are either extinct or very restricted in habitat. However, it still goes on in a few places and can create problems of several kinds. Hybrids between dogs and American timber wolves – wolfdogs – are popular with owners who relish a challenge. Legal in some US states and illegal in others, they can (if genuine – many are fakes) be unpredictable and dangerous. The conservation of the wildcat in the United Kingdom and Europe is seriously hampered by interbreeding with domestic cats, which belong to the same species, *Felis silvestris*;

there may in fact be no completely 'pure' wildcats in Scotland and possibly as few as forty individuals that completely fit the blueprint for a wildcat. Wildcats are a European protected species, and hence it is a legal offence even to chase one away from its den. The hybrids, however, classed as feral, can be dispatched with impunity on the slimmest of pretexts – for example, if suspected of raiding game birds' nests. Nowadays this poses a problem for the wildcat, or at least for those who wish to preserve it as a 'pure' species.[16]

In the past the problem would have been the near-impossibility of keeping wild males away from semi-domesticated females. Imagine the likely situation somewhere in the Near East, say 7,000 years ago. Wildcats would have surrounded each small nucleus of semi-domesticated cats in widely separated villages and small towns. Every adult male wildcat would have been attracted to the domestic females and probably highly effective at competing with the 'soft' semi-domesticated males. Under such circumstances, protecting the integrity of the incipient domestic genome would have required active intervention on the part of the human inhabitants. And yet the DNA of today's domestic cats shows that it was separating as many as 10,000 years ago.

To preserve the difference between domestic and wild, some factor would have to prevent all newly domesticated animals from mating with their wild counterparts. Ten thousand years ago, physical barriers – walls, fences – would have been far less effective than they are today. That's not to say that people wouldn't have built them: only by keeping their domestic animals enclosed could they guarantee that their investment would pay off in terms of help with hunting, milk, eggs or meat. With no understanding of genetics, they would not have understood the risks posed by occasional incursions of wild males or temporary excursions of domestic females in heat, both of which would have driven domestication backward.

However, treatment of some domestic animals as pets, in the sense of keeping them close by at all times, might have significantly reduced the likelihood of their mating with a wild member of their species. This could have helped retain the integrity of the 'domestic' genes, allowing further selection for those traits that made the animals more suited to whatever function they might serve.

None of this is to say that our early forebears were necessarily pet keepers in the modern sense. Domesticated animals still predominantly played a functional role. Some would have been 'walking larders', while domestic carnivores such as cats and dogs would have kept the settlement clear of vermin. However, any close relationship between an individual animal and an owner would have helped to protect such companions from the unwelcome attentions of their wild counterparts – regardless of whether their human minders understood its consequences or not. People may not have fully recognized that the offspring of animals that they kept close and that mated only with other domesticated animals were easier to handle than those whose parents had been less carefully tended. A simple emotional reaction – an affectionate desire to take extra care of those individual animals that appeared to reciprocate that affection – might have sufficed to keep domestication on course.

Overall therefore, we have two plausible explanations for why our evolution may have favoured both an affection for and a facility with animals. Initially, from the moment that pet keeping began – we don't know precisely when, but presumably sometime in the Paleolithic – pets might have functioned as advertising: as an 'honest signal', by men of their wealth and/or their ability to be gentle and caring, and by young women of their ability to raise young. Then, with the advent of agriculture, a facility with animals would have both kick-started and, perhaps more importantly, sustained the gradual emergence of domestic animals genetically distinct from their wild counterparts. Groups of people who were 'good with animals', or at least contained some individuals who were, would have thrived, as would animals who were 'good with people'.

The human race's relationship with the animals around them was symbiotic. The animals gradually changed genetically to become more docile and tractable – domesticated. Our ancestors gained the advantages that accrued from keeping animals and evidently out-competed the more traditional hunter-gatherers around them: people who were naturally good at taking care of animals left more descendants than those who excelled at hunting them.

It is less clear whether the people who domesticated animals succeeded because they were genetically different from their neighbours in terms of empathy with animals, or whether the crucial differences were cultural. We can say that socially transmitted skills – culture – must have played the major part once domestication was well under way. Specifically, as animals became more domesticated and less able to fend for themselves, their care would have become more complex and demanding. Each generation must have transmitted to the next its accumulated knowledge of how best to provide this; otherwise the domesticated varieties would have failed to thrive, and domestication would (locally) have gone into reverse. Although skill in looking after specific animals would have become part of the local culture, however, the motivation to acquire and apply these skills would still have varied from person to person, influenced genetically, just as a person's expressed desire to play with a companion animal is today.

Once domestication had progressed to the point where people could move domestic animals away from areas inhabited by their wild counterparts, genetic factors would have played less of a part, and cultural practices would have become more important. Even in the Neolithic period, customs, more than genetic differences, would have profoundly influenced the spread of domesticated animals between neighbouring tribes. Presumably the advantages of possessing, say, a pack of hunting dogs or a herd of goats would have been self-evident to those who had yet to acquire them, though those who had performed the original domestications might actively have tried to restrict their spread. A little over 2,000 years ago, Roman occupiers reported that the ancient Egyptians had laws against exporting cats, for instance. While these claims may not be entirely accurate, certain groups certainly may have understood the advantages of domesticating animals and tried to keep their knowledge from neighbouring rivals. Even if the animals themselves were freely traded (or appropriated!), they could only have thrived if their new communities possessed the requisite husbandry skills (though this may not apply to the largely self-sufficient domestic cat).[17]

Nevertheless, individuals with a genetic facility with animals would always have had a small advantage over those without. Even in groups that acquired their domesticated animals ready-made, such people would have naturally gravitated towards taking care of the animals and acquired the specific knowledge necessary, thereby reinforcing the advantage of possessing the underlying genes. These genes cannot have been crucial, because they are not universal in the modern descendants of the early pastoralists (us!). Today, some people feel an affinity for animals; others don't. We know that a positive attitude towards pets runs in families, and while some imitation may come into play, the genetic similarity between family members provides an equally plausible explanation. Although one twin study has demonstrated that such genes almost certainly exist, we don't know how many there are or precisely what they do. Nonetheless, the degree of variation in terms of how individual people feel about pets means that such differences are probably ancient and have been subject to selection throughout the history of our species.

Pets are not parasites. The vast majority of pet owners know perfectly well that their animals are not children. Not only can they confirm this to anyone who asks, but their brains react to pictures of their pets (and therefore most likely to the pets themselves) differently than they do to images of children, and they speak to their animal companions differently than they do to children. The tiny minority of people who hoard large numbers of animals may well harbour the delusion that these creatures are their kids, but extrapolating their misapprehensions to all pet owners is grossly unfair and inaccurate. Some people do seem somewhat susceptible to the cute features of puppies and kittens, but if nothing else sustained pet keeping, surely far more people would abandon adult cats and dogs than actually do.

The bond that builds between owner and pet must run deeper than any facile fascination with cuteness; rather, it may have emerged as part of our evolution as a profoundly social species – social, that is, not only with our own kind but also with other species. Dogs transformed our hunting habits, and cats controlled vermin in our barns and larders; goats, horses and cattle gave us milk, and pigs, cattle and sheep provided meat and hides. It's a truism to say that we domesticated them: in a very real sense they also domesticated us, as pastoralists came to dominate the planet, forcing hunter-gatherers into habitats unsuited to domestic animals.

Thousands of years of animal husbandry have left their mark on the workings of our minds, and surely it's not too fanciful to suggest that they have likewise altered the genes that structure our brains. Those of our ancestors who understood animals would have prospered at the expense of those who could not. Thirty thousand years ago they would have been better hunters; 20,000 years ago they would have harnessed dogs' abilities to guard their homesteads and improve their detection of quarry; 8,000 years ago, their food stores, protected by cats, would have lasted, while those of their neighbours disappeared into the mouths of mice. People with the capacity to form relationships with individual animals would, probably more by accident than design, have protected them from mating with their wild counterparts, thereby preventing dilution of the genetic material that distinguished useful domestic stock with the genes of wild ancestors and thus preserving the blood lines that persist right through to the present day.

Those who carried the genes for a facility with animals may also have found favour because this skill acted as a proxy for probable good parenting. Today's city dwellers might find this risible, despite being but a few generations removed from those for whom a facility with animals would have meant the difference between sufficiency and starvation. Nowadays, apart perhaps from an isolated farmer, few people pick a partner based on his or her skill with animals. Not many generations ago, however, even in Western societies, parents and prospective in-laws played a much more decisive role in match-making and may well have taken such abilities into account.

Thus, in some parts of the world, our ancestors who were good with animals may have prospered at the expense of those who were less good. This advantage cannot always have been decisive, because even today individuals and societies vary a great deal in their regard for animals. Nevertheless, historically speaking, all societies have relied on animal products for nutrition and on the assistance of such animals as dogs, cats and horses. Our brains may have changed – evolved – to enable this, but even if they did not, individuals with little facility with animals would have tended not to prosper. The genes that enabled our Paleolithic ancestors to comprehend what the animals around them were thinking are most likely the same as those which now allow us to enjoy keeping animals among us.

Animals Maketh Man

T HE INTIMATE AND uniquely flexible association between humankind and other animals extends so far back into our prehistory that we will never be certain when it began. If current theories
of how the human mind evolved are correct, it may have developed
gradually, between around 100,000 years ago, with the first ritual
burials of animals, and around 30,000 years ago, when the earliest
zoomorphic art appeared.

The crucial innovation that made the human–animal bond possible seems to have been a new way of thinking, not just *about* animals but *with* them. Millions of years ago, our hominid ancestors
must have gradually evolved more and more refined ways of thinking
about animals, enabling them to exploit a wide range of types of animal food, from shellfish to antelope, and thereby allowing them to
exploit a wider range of habitats than before. And once symbolic
language had evolved, at around 100,000 years ago, our direct ancestors could also have communicated their thoughts about animals to
one another – as could our Neanderthal cousins, if speculation about
their ability to speak are correct.[1]

One crucial difference between *Homo sapiens* and *Homo neander-
thalensis*, however, may have been our ability to take the animal's
perspective, to imagine what it might be thinking and thereby
predict what it might do next: in essence, the anthropomorphiza-
tion of animals. We probably didn't do this very accurately even in
those days, since most of us don't do it very well now. Nevertheless,

anthropomorphism – thinking *for* animals – seems to have worked well enough for our ancestors that they gained a significant advantage over their competitors (including the Neanderthals). In time, *Homo sapiens* would become the dominant predator across the whole of the globe.

Just as we cannot be certain when the human mind evolved the ability to analyse the thoughts of other species, we will probably never be sure precisely when our ancestors began to keep pets. However, the near-universal practice among present-day Paleolithic societies of obtaining animals from the wild to keep as companions suggests that the habit is extremely ancient. Given that many of the animals adopted by such societies are from wild species, this habit was most probably widespread many thousands of years before any species were domesticated. Taming of young animals to serve as pets therefore constitutes the first phase of humankind's personal relationships with animals.

The second phase, the domestication of a few key species, began with the dog sometime between 30,000 and 15,000 years ago. This event differed from many of the subsequent domestications in that the animal's behaviour became of primary interest to its human keepers, who had presumably realized that they might exploit the dog's superior sense of smell and speed. There are competing theories as to the initial function of domestic dogs – some say hunting, some guarding, some companionship – but the archaeological record suggests that within a few thousand years we had, metaphorically speaking, harnessed them to all three roles, in addition to literally harnessing them to travois. That one species could fulfil so many purposes reflects not only the malleability of canine behaviour but also the flexibility of the human mind – the ability to recognize that dogs were not, so to speak, one-trick ponies.

Anthropomorphic thinking would have been essential to this stage of domestication. To be useful, dogs require training, which in turn requires an understanding of their minds – to a greater extent than for most other domestic animals, which we need merely to control. The ability to predict and control what a sheepdog will do next is more important than the ability to predict the next move of the sheep.

Most later domestications exploited animals for their physical attributes – meat, milk, hides, brute strength – rather than for their minds and senses. One other exception, the cat, we used for its hunting prowess and agility, which remain largely undiminished to this day. In contrast to dogs, however, we left cats largely to their own devices, for the very good reason that they work most effectively alone.

Individuals from all the domesticated species had the potential to become personal animals – pets – but an understanding of the individual animal's temperament and personality was crucial to the effectiveness of the relationship mainly for dogs (and some draught animals). We draw most pets from domesticated species, but most domesticated animals are not pets.

The origins of the third phase, the mass adoption of domestic animals primarily for their company, probably go back to the beginning of the second phase. Even at the earliest stages, a few privileged people almost certainly kept favoured dogs as pets. Economic emancipation, however, most likely led to the explosion of pet keeping during the twentieth century. Given that the Abrahamic religions have from time to time held a dim view of either pet keeping or one or another of the species involved, the modern trend towards secularism may also play a part, allowing pet owners to express their affection for their animal companions without fear of stricture.

Our pet-keeping habit may thus be almost as old as our species, rather than an emblem of modern decadence. For thousands of years of recorded history, most people were tied to the treadmill of agriculture (as many still are today) – so pet keeping as practised today was far less widespread then than it is now. But it has inhabited the human psyche for many millennia.

Our history suggests that keeping pets is perfectly normal when circumstances allow. Recently, however, support for the benefits of doing so – perhaps most notably the notion that they can help us live longer – has spread with remarkably unbridled enthusiasm.

This phenomenon, far more than the impulse to keep pets in the first place, requires explanation.

Many of the supposed advantages that pets confer on their owners are either dubious or unproven. The supposed health-enhancing properties of pets increasingly appear largely illusory, resting on the selective reporting of studies seeming to show some kind of positive connection between well-being and pet ownership. There may be a small effect for dog owners, but more plausible causes than the dog's company exist. It's more logical to propose that the link works the opposite way around – that people who are unwell tend not to get a dog out of legitimate concern that they might not be able to give it the lifestyle it deserves. Indeed, they may plausibly get cats instead, which might explain why the few studies that have included cats tend to associate their ownership with poor physical or mental health.

Any genuine health benefits arising from owning dogs may result from the need to exercise them: walking a dog in the countryside (sans mobile phone) inevitably exposes the walker to the better-documented benefits of green exercise (see nearby illustration). Only if energetic enough, however, will the walk likely improve the owner's cardiovascular fitness. Of course, those with enough motivation could get exactly the same benefits from a jog around the park, without the potential hassle of having to call the dog away from over-protective owners who think their dog is above meeting others of its own kind. But I know from experience that the most reliable prompt for a twice-daily walk is a dog charging around the hallway with his lead in his mouth and his tail in overdrive!

The most recent studies have not confirmed the supposed link between pets and longevity, so it is not at all clear that pets provide the equivalent of social support – once a popular notion among anthrozoologists. People with strong social networks are generally healthier and live longer than individuals who are more isolated: this remarkably wide-ranging effect extends to all manner of ailments, including cancers, cardiovascular disease, some infectious diseases, rheumatoid arthritis and schizophrenia, in addition to the more expected link to

Any health benefits that result from ownership of a dog we can plausibly ascribe to the relaxing effects of exposure to green spaces.

loneliness, depression and suicide. Although scientists have not yet agreed fully on why sociability staves off disease, the evidence that it does is substantial.

Whatever the aspect of human company that makes people healthier, pets don't seem to provide it. Pets do demonstrably reduce the physiological effects of mildly stressful situations, but stresses that pet ownership creates probably cancel out these benefits. Thus, pet keeping may have no net effect on the cumulative effects of stress as far as the owner is concerned: the good times balance out the hassles, but no more.

The animal-assisted therapy (AAT) industry rests on the idea that animals (especially dogs) confer a touch of magic for people with impaired health or well-being. But again, that the animals themselves have any such effect is not at all clear. Some instances may involve a sizeable measure of placebo or wishful thinking, as exemplified by the dubious practice of dolphin therapy. Where AAT is genuinely successful, the simple presence of the animal has most likely induced people to interact more fully with each other and to trust each other more. The animal is an enabler of *human* connection – not itself the determinant of the interaction. There are a few indisputable exceptions – notably, the bond that can form between some children on the autism spectrum and their animals – but it is this 'bridging' element that may be key to understanding much of modern pet keeping.

We are also much less certain than formerly that a 'humane education' that includes looking after animals inculcates children with empathy for their fellow human beings, despite wide promotion of this outcome. Research has shown that variations in how sympathetic people are towards animals bears very little relation to how empathetic they are towards their fellow humans. Certainly the supposed progression from cruelty to animals in childhood to violence towards humans later in life has not stood up to scrutiny. The two may have some predisposing factors in common, but these have little relevance to the comprehension of everyday pet keeping.

The recent onslaught of pro-pet thinking possibly arose out of twentieth-century threats to pet keeping emanating from the medical profession and other authorities. As the numbers of pet cats and dogs increased, dire warnings of the risks they posed to society surfaced. Cats became scapegoats for suicides and cases of schizophrenia, the supposed link being the toxoplasmosis parasite. As the numbers of dogs exercised in city parks grew, mounting public disgust forced authorities to introduce poop-scoop laws. More recently, conservationists have condemned cats as heartless murderers of innocent wildlife. Some had demonized certain dog breeds as a danger to the public due to their supposedly uncontrollable aggression. Yet the popularity of pet keeping remains undimmed.

Owners become so fond of their animal companions that they seem willing to accept any 'fact' that will countermand the negative attitudes that surface in the media from time to time. What greater justification for keeping a cat than the belief that it is every bit as good at staving off loneliness as a human companion would be? The city dog owner searching for a bin, plastic bag of still-warm faeces dangling from one slightly outstretched hand, can rest assured that the walk he is about to complete has saved him from a heart attack. Owners can bask in the reflected glory of those pets that seemingly perform miracles – reconnecting autistic children with the world or releasing an elderly person from the clutches of dementia – and so have little incentive to probe these claims for plausibility. It's human nature to prioritize good news over bad when it reflects on those we love.

EVEN IF WE could prove all the proposed health benefits of pet keeping, they cannot explain why we keep pets in the first place. Undoubtedly a few people have taken on a pet based on claims made on the Internet or in the popular press that by so doing they will instantly make new friends, become healthier, and feel less stressed, but this motivation doesn't factor into most people's decision to acquire an animal companion. The ailments that pets supposedly excel at alleviating – heart disease, loneliness, overwork, allergies – are the product of modern lifestyles and so cannot have caused the pet-keeping habit to evolve tens of thousands of years ago.

So why, then, do we keep animals as companions? To fully understand this phenomenon, we must revisit four characteristics of human nature that help to bind us to our pets. Three of these have plausible origins in prehuman behaviour; the fourth probably did not emerge until part way through the evolution of our own species.

The most ancient of these is the 'cute response', which evolved, hundreds of millions of years ago to ensure that mammalian mothers (and in some species, fathers) took special care of their offspring. Among humans, it still very much fulfils that purpose with regard to our own infants. But some components of the cute response in our brains also respond to baby or babyish animals. Not all, certainly: some parts of the brain only fire when the trigger is genuinely human. Thus, at some point in our evolution, the cute response bifurcated:

its core was retained as essential to ensuring that we cared for our own infants, while a partly separate system that directs us to care for animals appropriated some elements.

This bifurcation clearly served a purpose; otherwise the trait would not have survived for millennia. If some animals distract us from human infants or cause us to direct some resources away from them, it's not because our brains have somehow tricked us into transferring our affections. The idea that pets are hijacking care that we should direct towards other humans – be it by inhibiting potential mothers from having children or absorbing funds that should go to famine relief – presupposes that people don't know the difference between a pet and a fellow human being. That people are fully cognizant of the difference, however, is evident at every level, from patterns of brain activity, to the different language used to address pets versus babies, to consciously expressed opinion. Cuteness makes us attentive, but its charms do not blinker us. (Here, I'm referring specifically to the appeal of baby-like features. Pet owners have an unfortunate tendency to label as 'cute' anything they like about their own animal – hence the apparent increase in their ratings of 'cuteness' as the relationship develops.)

The evolutionary origins of the system that responds to animal 'cuteness' require some explanation. Why do animals that are manifestly not human infants nevertheless trigger our cute response, when care directed towards other species could potentially disadvantage our own offspring? Why are we not much more selective, given that we can easily distinguish between a kitten and a human baby? It's possible that we're simply too gullible, but I believe that the appeal of cute animals, far from being an error, is a consequence of our evolutionary history, signifying the benefits our distant ancestors accrued from incorporating animals into their social circles.

Almost as ancient, because it occurs in all primates, is the bonding power of stroking. Despite its ubiquity, touch is the one element of pet keeping that hasn't received much attention from scientists, but it is probably more important than cuteness in sustaining relationships with pets. We love to stroke our cats and dogs. Explaining in evolutionary terms the pleasure we take in feeling soft fur under our fingers is not straightforward, given that hair covers such a small proportion of our own bodies. Yet, whereas all the primates are hairy, our ancestors evolved largely naked skin some 1.6 million years ago, probably to allow them to run long distances without overheating: sweat evaporating from hair cools only the hair, whereas sweat evaporating from skin cools the body. But this advance would have sacrificed an essential element of primate social life: the mutual grooming that sustains relationships between members of the same troupe. An unresolved question posed by the evolutionary approach to psychology has to do with how our ancestors managed to maintain the cohesion of their groups without this apparently essential mechanism before we had evolved its eventual replacement, language. We might therefore regard the pleasure we get from stroking our dogs and cats as simply a throwback to our primate past. This may indeed be its origin but cannot explain why it is still appealing today. In short, why did furry skin not lose its appeal when our own skin became smooth?[2]

Other aspects of pets' tactile appeal also require explanation. We don't just find touching them pleasurable; we generally find it relaxing. The accompanying changes in our hormone levels must have some meaning, setting aside the myth that stroking a dog or cat calms us to the extent that we end up living longer. It's not enough to say that we like stroking our pets because it causes the release of dopamine and endorphins in our brains: that just gives the experience a physiological basis. The crucial question is, why does this mechanism still exist, hundreds of thousands of years after it should have died out, once our ancestors lost their furry coats? Why do we still prefer to touch hairy skin rather than smooth, when hairlessness is one of our defining characteristics? Why are hairless dogs and cats generally regarded as freaks, when they are in that single sense more like us? Why does a few minutes of caressing a furry surface, even an inanimate one like a plush toy, have the potential to radically improve our mood? The answer may be that we preserved this instinct because, even after it became irrelevant in our largely hairless human social groups, it continued to serve us in our ability to cement bonds with our domestic animals. These bonds promoted their survival and, by extension, our own.

A third facet of human nature that enables our relationship with pets is the reward that many of us derive from caring for young animals. All mammals, of course, take care of their own infants until they are capable of fending for themselves. Humans are unique in finding enjoyment in nurturing the young of other species. Even inanimate objects, such as Tamagotchi and AIBOs, can trigger similar feelings, presumably because they are responsive, or at least produce the illusion of being so. As for the 'cute response', we need to find an explanation for this apparently profligate use of our time, energy and resources, which in narrow genetic terms we could better devote to our fellow humans. Other species focus their attention on their own kind and don't waste their energies keeping pets. Why only us? And why women and children, especially? If we reject the idea that pets are simply parasites, our ancestors must have accrued some advantage from this affection for animals; otherwise it would have died out. That advantage was, most plausibly, a capacity to domesticate animals, permanently separate the tame from the wild, and thereby stabilize their supply of animal protein.

In modern Paleolithic societies women and children usually took responsibility for caring for the pets, even though men had usually retrieved them from the wild. This may be a genuine sex difference, rather than simply custom. Even today, women typically express deeper affection for animals, although there is a great deal of overlap between the sexes. Given the likely involvement of oxytocin in the bond between women (especially) and their pets, such gender differences may have a physiological basis – for example, the reduction in loneliness that pets bring to late-middle-aged women, but not men, may stem from the greater hormonal response women experience from taking care of their animals day after day. The positive feedback that this gives them is crucial to sustaining and deepening the relationship between women (especially) and their pets. We may therefore be reasonably sure that women, more so than men, preserved the precious genes that promoted tameness in the early stages of animal domestications.[3]

Our fourth evolved mechanism for relating to pets is our ability to anthropomorphize, to try to get inside the minds of the animals around us. This trait may well have first arisen with the merging of our social, natural history and general intelligences, which, though difficult to date precisely, appears to have occurred gradually between 100,000 and 30,000 years ago. Partly derived from brain mechanisms that analyse the behaviour of other humans, it initially evolved as a way of guessing the intentions of predators and prey, thereby conferring a considerable survival advantage on those humans who were best at it. Once established in our brains, it would have helped us predict the behaviour of captured animals also, enabling pet keeping to thrive. It is an imperfect mechanism for understanding what animals are really thinking, since their minds, not to mention their sensory abilities, differ substantially from our own. Nevertheless, anthropomorphism appears to have been adequate to confer a significant advantage to our hunter-gatherer ancestors, since it is a universal of human behaviour even today.

Anthropomorphic thinking presumably lies behind the somewhat lazy truism that pets are 'one of the family'. Our behaviour towards our pets shows that, for most of us, they inhabit a unique space somewhere between kin and chattel, not quite human but not quite possession either. Within that space we anthropomorphize them more intensely than we do any equivalent wild animal, as if wishing to draw them closer to being 'little people' and denying their wild origins – but never quite succeeding.

Cuteness may be paramount in the initial appeal of a particular animal, but the rewards gained from talking to, caring for and stroking them sustain the relationship between owner and pet. At one level, these behaviours explain much of the unique character of pet keeping; however, they all beg the question, why do we react in these ways towards members of other species? Did our ancestors benefit from such apparent squandering of resources to the extent that the ability to keep pets became almost a universal characteristic of humankind? Taken together, this suite of traits – the appeal of cuteness and soft fur, the sense of reward derived from caring, and a desire and facility for understanding what animals are thinking – put some of our ancestors on the winning side, not once but twice: first when pet keeping became an emblem of a careful and caring person and again, later, when pastoralism and then agriculture supplanted hunting and gathering as our species' defining way of life, inhibiting us from wiping out our newly domesticated animals when we were starving (see illustration on p. 303).

THE WAY IN which such behaviour became part of human nature must lie in the evolution of our own species, which, of course, is not an easy matter to reconstruct. However, I propose a sequence of evolutionary processes that could have led to certain of our ancestors developing a powerful affinity for animals. First, the development of anthropomorphic thinking allowed us to become smart hunters, using more sophisticated techniques than the crude methods of our nearest rivals, the Neanderthals and the Denisovans. Then a facility for taking care of animals caught in the wild might have become a factor in the selection of marriage partners, initially as a proxy for skill at raising human offspring, since raising animals at that time would have provided little advantage over simply hunting them. Next, a capacity to feel affection for animals must logically have been essential to the domestication of wild animals, protecting the precious genetic material that allowed certain animals to live in human settlements from dilution through cross-breeding with their wild counterparts. Since societies with domestic animals evidently thrived at the expense of those with none, any human genes favouring such emotional ties would have spread at the expense of those that favoured a more exploitative approach (bluntly, eating the breeding stock). Furthermore, once domestication had got under way, individuals who were good with animals would have stood out as such, and this facility may eventually have factored into decisions about who would marry whom, inhering in the traditions of those societies transitioning from hunter-gatherer to pastoralist. In this way changes in local culture would have reinforced preexisting genetic differences: individuals genetically predisposed to an aptitude with animals could exploit their talent fully by learning the details of husbandry (which they could not have inherited genetically) from those around them. This scenario may initially seem far-fetched, but it neatly explains why, to this day, a person with a well-behaved animal by his or her side seems more trustworthy and approachable. Even now, many of us instinctively seem to like people who appear good with animals.

We can also explain the evident fascination that animals hold for children in terms of the expression of genes that affect affection and care for other creatures. If the ability to take care of animals was a factor in prospective in-laws' decision to marry off their son to a particular girl, overt signs of affection for animals would logically be strongest during puberty. It is highly unlikely that any gene favouring a facility with young animals would have acted directly on behaviour, if for no other reason that that each species requires a subtly different kind of care, and different species would have been available to each society (as today), depending upon where they lived. The greater likelihood is that some of the genes concerned conferred a curiosity about animals and an ability to learn how to take care of them, capacities that we still observe (and, rightly, encourage) in our children today.[4]

We also seem predisposed to accept much of what we hear about animals, explaining the popularity of the notion that having a pet at home will automatically improve our health or make our children better citizens. It's almost as though anything that triggers the 'friendly-animal-responsive' parts of our brain, be it a direct experience or a photo accompanying an article, makes us feel good and thus less sceptical. Advertisers certainly know this, even if unaware of the underlying brain mechanisms that allow them to sell products from toilet tissue to mobile phones using images of cute dogs and cats.

SINCE KEEPING OUR domestic animals physically close to us must have been crucial to domestication, the pleasure gained from touching them may have contributed to maintaining the bond be-tween owner and 'pet', both way back in the Paleolithic, when pets primarily advertised gentle behaviour, and subsequently as an aid to domestication. Nowadays most of us must restrict our tactile contact to our pet cats and dogs, but the fur of some other sociable domes-tic animals may be just as appealing to those with an opportunity to touch them. As country-dwelling Twitterati cat impersonator Tom Cox once posted, 'Go out. Find a lovely sheep. Scratch its head. Go on. You know you want to. The sheep knows you want to. It's OK.'[5] This effect would not have applied equally to all domestic species – they're not all equally cuddly, and some have feathers, not fur – but for dogs and then cats, it may have been crucial. Wolves' and wild-cats' pelts may be just as tactile as those of their domestic counter-parts, but their bearers' disinclination to come within reach of a human hand would have ruled them out – until they'd been turned into garments, that is. Meanwhile the Mesolithic owner absentmind-edly rubbing his dog between her ears would have unwittingly con-tributed to keeping her from pairing up with a wolf, thereby ensuring that her puppies were just as appealing as she or (better) even more so. Our ancestors' evolution of hairlessness freed our predilection for the feel of soft, furry surfaces from its original role in social bonding, allowing it to become part of our relationship with other mammals.

Anthropomorphism would have provided an advantage during all these stages: the ability to get inside an animal's head, even if imper-fectly, would have given those who had it an advantage both when hunting and when caring for personal animals. All stages in the cul-tural development of modern humans would have favoured the 'cute response', the ability to derive satisfaction from taking care of an an-imal, and a tendency to take pleasure in stroking furry skin – perhaps most crucially during the early domestication of the dog. Indeed, it may not be too fanciful to suggest that hunter-gatherer pet keeping became a habit because it enabled people to indulge their ancient primate urges to form bonds through grooming.

In this scenario, all the relevant changes to our brains could have occurred long before pet keeping became the norm and thus well

FOUR KEY EVENTS IN THE EVOLUTION OF *Homo sapiens* THAT ACCOUNT FOR THE MODERN POPULARITY OF PET-KEEPING

3,000,000 - 1,000,000 years ago (approx)

Our hominid ancestors become hairless, but retain their fondness for stroking fur

50,000

Areas of the brain that analyse human and animal behaviour connect, permitting anthropomorphism

30,000

Young women who demonstrate an ability to care for animals (as a proxy for children) are preferred as brides

20,000

Taboos against killing and eating personal animals allow domestication to proceed uninterrupted by famine

Today

Four key events in the evolution of *Homo sapiens* that account for the modern popularity of pet keeping.

before any domestication of animals took place. Subsequently, the advantages gained by being good with animals would have caused these genes to become more widespread. It's also possible – though not essential to my argument – that further changes in how our minds work occurred during the past 10,000 years as a consequence of our domestication of animals. We know that our exploitation of animals for their milk has affected the genes that program our digestive systems, and an increasing body of evidence suggests that our brains have continued to evolve throughout the history of our species and may still be changing today. It's not entirely out of the question that genes that only provided an advantage after we domesticated animals have now joined the prehistoric ones that predisposed our ancestors to adopt wild animals.[6]

WE CAN THUS view modern pet keeping as the contemporary expression of an ancient connection with animals, one that proved very useful to many of our ancestors at different stages of our evolutionary history and thus became an intrinsic part of human nature. Not universally so, however: even today there is still some evidence for the remains of the variation between individuals that was the raw material for such an evolution. Humans have evidently prospered by other means than a facility with animals – for example, in societies that thrived by exploiting others, as the 16 million male descendants of the Mongol warrior Genghis Khan demonstrate. For example, the twin study showing that middle-aged men's fondness for playing with animals is heritable demonstrates that while some individuals are genetically predisposed to like animals, others are predisposed not to. At some point in the past, the genes that allow their bearers to comprehend and feel affection for animals would have been rarer than they are now. Those who happened to carry them thrived at the expense of those who did not, but not to the point where everyone became equally adept.

Those very genetic differences that helped some of our ancestors to domesticate animals while others did not may still underlie some of the disagreements between today's pet lovers and their detractors. Social commentator Laura Marcus (see the introduction) reported that her parents 'believed we humans had no right to own other animals. I agreed with them then, and still do'. We would expect precisely this if an attraction to animals were heritable; we would also expect this trait not to have become universal, even in the pet-obsessed West. Although the heritability could be entirely cultural – at its simplest, children copying their parents – I'm inclined to believe that there must also be a genetic component. Children don't acquire their personalities wholesale from observing their parents; they inherit about half (genetically) and then acquire the rest, in poorly understood and probably highly individualistic, even haphazard ways, from their experiences as they mature. Their approach to animals might similarly derive from a blend of genetics and experience.[7]

Put another way, if a love for animals so advantaged our ancestors, why don't we all feel it? One answer is probably that, historically, there was more than one way to acquire domestic animals and thereby gain an advantage over societies still reliant on hunting. One was to domesticate them oneself, but it might have been equally advantageous in the long run to acquire, by fair means or foul, any animals the neighbours had, along with the knowledge of how to look after them. The flipside of a propensity to lavish affection on animals is that one might spend that time more gainfully on a task that occupies different brain circuits – for example, perfecting one's arsenal of weapons to defeat the goat herders over the hill. Thus we might reasonably suppose that some societies evolved a higher frequency of the animal-loving genes than others. The proportion of each within a given area might also depend on the wild animals available locally for domestication. For example, there were no grey wolves in Africa to domesticate into dogs, and the Ethiopian wolf and African wild dog appear biologically unsuited to such a transition. A lack of selection for animal-loving genes may explain why the Kiembu people of Kenya have no word for 'pet': they may keep dogs as guards but never play with them or let them into their homes. Such biological constraints, acting alongside a multiplicity of cultural factors, might explain why pet keeping varies so much around the globe. The much more rapid variation in the West, where breeds and species drop in and out of fashion almost yearly, can give the impression that pet keeping is a meme – a self-replicating idea that travels from one human brain to another – powerful enough to override any genetic predisposition. These rapid changes in the detail of pet keeping can, but should not, obscure the underlying human need to form affectionate relationships with real live animals. Without it, we would already be moving towards the convenience of robot pets as rapidly as clean electricity replaced dirty steam on our railways.[8]

Like personality, affection for animals is likely influenced not by one but by hundreds or even thousands of genes. Research into the mechanisms whereby the human genome affects personality is still in its infancy, partly because there are so many genes involved but also because the effects of every single one of those genes varies according to each person's individual experiences. What's worse for scientists trying to pin down the effects of genes is that our big brains allow us to make decisions that determine the experiences we are likely to encounter: for example, extroverts move from rural areas into towns so they can meet other extroverts and are replaced by introverts seeking peace and quiet. However, one recent study shows how we might eventually tease out the effects of single genes. The ability to understand what other people are thinking – theory of mind (which is distinct from but may overlap with the ability to guess what animals are thinking) – develops in children at different rates. One factor affecting that rate is how sensitive to others the child's mother is; another is a gene that produces an oxytocin receptor in the brain, which in turn probably affects how the child's developing brain wires itself up. These two interact, so that children with the 'best' version of the gene (best, that is, for developing theory of mind; it might be the worst for something else, since oxytocin has so many effects) respond disproportionately well to their mother's example. Interactions like these might one day explain why a love of animals runs in families.[9]

OUR BRAINS HAVE evolved ways of enabling us to enter into personal, affectionate relationships with animals because our ancestors who possessed the genes that generate those mechanisms prospered. Pets allow us to have experiences and express behaviours once crucial to our survival (see illustration). Our human nature is difficult for us to ignore, so we persist in seeking outlets for many of our evolved behaviours, some of which fit with modern conditions better than others. We no longer engage in skirmishes with neighbouring tribes, plundering their crops and making off with their women and children. Instead, we have evolved (culturally) a whole raft of formalized team sports. We can (partly) fool our sweet tooth by choosing low-calorie alternatives, which might by now be outselling the seven-spoonfuls-of-sugar originals were it not for health scares over artificial sweeteners. We need to find work-arounds for some of our Pleistocene instincts because they are embedded in our genes. The company of a dog or cat is the pleasurable expression of a set of

Hunting with animals allows reconnection
with our ancestors' reliance on them.

Pleistocene patterns of behaviour that may serve little function today, but that does not mean we should try to extirpate them from human nature.

Keeping a pet is a perfectly natural thing for people to do – at least for those of us who are predisposed to do so. There is no evidence that we do ourselves harm, psychologically or otherwise, by enjoying the company of an animal. Science has confirmed that dogs and cats well socialized to people do feel genuine affection for their owners, so the 'unconditional love' that owners claim they provide is real enough, albeit occasionally misinterpreted. Rather than as 'unconditional', which smacks of some kind of contract, I prefer to think of pets' affection as 'uncomplicated'. They do not agonize for hours about how their relationship with us is going: 'Does my owner love me as much as he did yesterday?' 'Does he love me more or less than he loves his kids?' Such thoughts are beyond these animals' comprehension. The welcome that they give us when we come home is genuine and should be enjoyed for what it is. As social animals ourselves, we crave company even when our circumstances mean that some of the time we have to be alone. Simply having another living thing around the home – the click of the dog's toenails on the floor, the brief stirring of a half-sleeping cat as we enter the room – is deeply reassuring in itself. Pets are enjoyable (most of the time!) and should be a straightforward pleasure, unburdened by feelings of guilt that they might somehow be diverting us from a 'proper' social life. Pets may not have the power to heal us that some claim on their behalf, but they do bring us a sense of well-being. A recent international survey, 'The Rest Test', found that a quarter of respondents chose 'being with animals' as one of the three main ways they chose to relax.[10]

Pets make their owners happy (on balance). They may not make their owners any healthier, but they don't pose any significant risk to health either, at least in the West where zoonotic diseases are under control. Dogs and some cats can act as bridges between people and thereby enrich their lives. Many pet owners confess that without an animal companion, their lives would feel incomplete. This is not an admission of some kind of social inadequacy; it is the expression of a perfectly natural and long-standing human trait. For most owners, the inclusion of animals in their social networks enhances their relationships with other humans, rather than, as some would have it, providing an inferior substitute.

THOSE WHO DECRY pet keeping on the grounds that it has a negative effect on the environment would do well to note that in terms of their psychology, these apparently opposing perspectives may have more in common than they disagree about. In reality, most pet owners attempt to balance their love of animals with respect for the environment. Many of those who feed garden birds also own cats (in practice, it's not difficult to keep the birds safe from the cats). Dog walkers appreciate (and benefit from) the natural features of the places where they go to exercise their pets. Anecdotally, many of the best supporters of wildlife charities are pet owners.

Recent research suggests that people who have strong relationships with their pets also have positive attitudes towards wild animals and indeed the natural world as a whole; some refer to this as the 'pets as ambassadors' hypothesis. It's tempting to see these connections as dividing people into those who regard themselves as part of the natural world, loving pets and wildlife alike, and those who conform to the anthropocentric notion of man as separate from and superior to nature. The separation between 'pet' and 'wild' may be more apparent than real.[11]

If we are to ensure that the Earth remains habitable, we must convince everyone that the natural world is worth saving, and to do so we will have to reverse mankind's ever-increasing detachment from it. Pets could so easily be part of that solution, not part of the problem, provided that pet owners fully realize that animals, rather than being simply commodities, are connected by descent to the world of nature. The emotional connection that owners feel for their animal companions is not so dissimilar from the emotional connection that conservationists feel for the natural world.

Education must be the key. There is truth in the idea that pets teach children how to care – perhaps not how to care about each other but simply how to care for animals – and thereby about many of the fundamental realities of life. Now that most children in the West do not encounter animals as part of their everyday routines (apart from their pets), it seems a terrible waste not to use their instinctive fascination with animals to teach them some basic biology.

The very week I first drafted this chapter, my five-year-old grand-daughter Beatrice's class at school had been monitoring an incubator full of hen's eggs. When I delivered her to school in the morning, and she saw the eight newly hatched chicks and eight broken eggshells, the excitement among her classmates was almost palpable. More-over the novelty of the experience evidently helped her remember much of what her teacher had told the class about what they had wit-nessed: 'The chic hach owt ov the egg and hee had a egg tooth,' she wrote that evening (see illustration). Whether or not she grows up to have a cat or dog of her own, a fascination with animals was kin-dled in Beatrice's mind that day. Her generation will have to stabilize the ecology of the planet for their own survival. Why would they want to do this without knowing the reality of animals, both pets and wild?

A child's fascination with animal life.

Acknowledgements

Throughout the planning and writing of this book, I have been conscious of standing on the shoulders of two giants in the field, Professor James Serpell of the University of Pennsylvania and Professor Hal Herzog of Western Carolina University, not only through the many illuminating conversations I have had with both over the past three decades but also through their seminal and very accessible books: Serpell's *In the Company of Animals* (without which there would probably be no science of anthrozoology) and Herzog's *Some We Love, Some We Hate, Some We Eat: Why It's So Hard to Think Straight About Animals*. More widely, I owe a debt of gratitude to all those, too numerous to mention individually, who have helped to build the International Society for Anthrozoology into the vibrant and successful organization that it is today. Without its conferences and its journal, *Anthrozoös*, gathering the raw material for this book would have been so much more difficult (and not half so much fun!).

My academic colleagues at the University of Bristol's vet school have provided a very valuable sounding board for many of the ideas that appear in this book (and a few that, for good reason, do not!), and I should especially thank Dr Elizabeth Paul for sharing her insights into the psychology of pets and their humans and for commenting on some of the chapters. Many thanks to Dr Barbara Schöning for translating Paul Spindler's paper from the original German, and to Dr Akaya Matsumoto (Miura) for checking the Japanese script on p. 141. I am also grateful to Professors Henry Buller of Exeter University and Richard Bennett of the University of Reading for inviting me to participate in a series of workshops, funded by the United Kingdom's Economic and Social Research Council, that introduced

311

me to the work of the many social scientists studying human–animal interactions.

I am also hugely indebted to Professor Nickie Charles of Warwick University for sharing her unpublished work and for allowing me to reproduce quotes from her Mass Observation study of pet owners, as well as for critiquing the chapters in which those quotes appear. Professor Steven Mithen of Reading University may be surprised to find his book *The Prehistory of the Mind* providing much of the inspiration for Chapter 9.

I found I had so much to say about anthrozoology that condensing it all down into a book of manageable length has been quite a challenge. I am hugely indebted to my agent Patrick Walsh and my editors at Basic Books and Allen Lane, Lara Heimert, Jen Kelland and Tom Penn, for helping me to remain focused and eliminating the quirks in my academically trained prose.

This is my third book that my old friend Alan Peters has illustrated; it was also the most demanding, since every picture I invited him to draw was a one-off, involving many kinds of animal, from camel to falcon, and many varieties of human. He has, as ever, done my ideas proud.

Finally, I must thank my wife and family for keeping me sane during the past few years of writing and especially my granddaughter Beatrice for allowing me to include one of her earliest pieces of writing in the final chapter.

Further Reading

The two essential reads for anyone interested in the science of human–animal interactions are James Serpell's *In the Company of Animals: A Study of Human–Animal Relationships*, 2nd edn (Cambridge: Cambridge University Press, 1996) and Hal Herzog's *Some We Love, Some We Hate, Some We Eat: Why It's So Hard to Think Straight About Animals* (New York: HarperCollins, 2011). As a new academic subject, anthrozoology has few standard texts: Samantha Hurn's *Humans and Other Animals: Cross-Cultural Perspectives on Human–Animal Interactions* (London: Pluto Press, 2012) is the best that I've come across. The ethical issues surrounding pet keeping can be complex, but Peter Sandøe, Sandra Corr, and Clare Palmer have dissected them expertly in their illuminating book *Companion Animal Ethics* (New York: Wiley, 2016).

Those seeking both facts and opinions will find an abundance of both in Marc Bekoff's four edited volumes of the *Encyclopedia of Human–Animal Relationships: A Global Exploration of Our Connections with Animals* (Westport, CT: Greenwood Press, 2007). For a historical perspective, Linda Kaloff has written a one-volume account, *Looking at Animals in Human History* (London: Reaktion Books, 2007) and edited the six-volume set *A Cultural History of Animals* (Oxford, UK: Berg, 2007). Brian Fagan's *The Intimate Bond: How Animals Shaped Human History* (New York: Bloomsbury Press, 2015), provides a detailed anthropological account of animal domestications. Donna Haraway's *When Species Meet* (Minneapolis: University of Minnesota Press, 2008) is widely discussed among social scientists. For an intimate account of what it might be like to be an animal, I greatly enjoyed Charles Foster's *Being a Beast* (London: Profile Books, 2016).

Many other books discuss human–animal relationships, including several focused on animal rights. Those I've used as sources are listed in the notes to each chapter.

Notes

Preface

1. The history of ISAZ can be found at 'The Origins of ISAZ', ISAZ, http://www.isaz.net/isaz/history.

Conventions

1. For a vet's rebuttal of 'pet guardian', see Animal Health Institute, 'Pet Owner or Guardian?', AVMA, November 2005, https://www.avma.org/Advocacy/StateAndLocal/Pages/owner-guardian-ahi.aspx.

2. For a typical ethicists' view, see Andrew Linzey and Priscilla Cohn, 'Terms of Discourse', *Journal of Animal Ethics* 1 (2011): vii–ix; see also Kathy Matheson, 'Pet? Companion Animal? Ethicists Say Term Matters', Phys. org, 4 May 2011, http://phys.org/news/2011-05-pet-companion-animal -ethicists-term.html.

Introduction

1. Natalie O'Neill, 'Woman Kicked Off Flight After Pet Pig Stinks Up Plane', *New York Post*, 28 November 2014, http://nypost.com/2014/11/28 /woman-kicked-off-flight-after-hiding-pet-pig-in-a-duffle-bag.

2. 'Therapy Piglet Could Change 6-Year-Old's Life, Mom Says', KCCI, 19 February 2015, http://www.kcci.com/news/therapy-piglet-could-change -6yearolds-life-mom-says/31352204.

3. Abraham Rosman and Paula G. Rubel, 'Stalking the Wild Pig: Hunting and Horticulture in Papua New Guinea', in *Farmers as Hunters: The Implications of Sedentism*, ed. Susan Kent, 27–38 (Cambridge: Cambridge University Press, 1989).

4. The male piglets are sensibly all castrated, since the adult males are dangerous: feral males living in the forest sire each generation of piglets.

5. For accounts of pet keeping in the United Kingdom and United States, see Keith Thomas, *Man and the Natural World: Changing Attitudes in England, 1500–1800* (London: Penguin Press, 1984); Katherine C. Grier, *Pets in America: A History* (Chapel Hill: University of North Carolina Press, 2006).

6. Anthony Podberscek, 'Good to Pet and Eat: The Keeping and Consuming of Dogs and Cats in South Korea', *Journal of Social Issues* 65 (2009): 615–32.

7. This quotation comes from a submission to the Mass Observation Project, which has been collecting anonymized opinions on specific topics from a panel of over five hundred UK correspondents since 1927. See Nickie Charles, '"Animals Just Love You as You Are": Experiencing Kinship Across the Species Barrier', *Sociology* 48 (2014): 715–30.

8. From 'Pet Population 2014', PFMA, http://www.pfma.org.uk/pet -population-2014. Fish are the most numerous animals kept in homes – about 20 million in the United Kingdom and 90 million in the United States – although in both countries these are minority interests, accounting for only about 10 per cent of households.

9. 'A List of the Dogs Banned in Beijing', *New York Times*, http://www .nytimes.com/2013/06/23/world/asia/a-list-of-the-dogs-banned-in-beijing .html.

10. For statistics on spending on pets, see 'How We Spent £6bn on Our Pets This Year', *Telegraph*, http://www.telegraph.co.uk/finance/newsbysector /retailandconsumer/11151275/How-we-spent-6bn-on-our-pets-this-year .html; Associated Press, 'Americans Spent $58 Billion on Pampering, Feeding and Protecting Pets in 2014 . . . Nearly $3 Billion More Than 2013', *Daily Mail*, 5 March 2015, http://www.dailymail.co.uk/news/article-2981501 /Americans-spent-58-billion-pamper-protect-pets-2014.html.

11. Rod Liddle, *Sun*, 13 June 2013, 13.

12. Hilit Finklet and Joseph Terkel, 'Dichotomy in the Emotional Approaches of Caretakers of Free-Roaming Cats in Urban Feeding Groups: Findings from In-Depth Interviews', *Anthrozoös* 24 (2011): 203–18.

13. Adapted from Stephen Kellert, 'American Attitudes Towards and Knowledge of Animals: An Update', *International Journal for the Study of Animal Problems* 1 (1980): 87–119.

14. 'Cat Quotes by Author', The Great Cat, http://www.thegreatcat.org /stories-poems-and-quotations/cat-quotes-by-author.

15. 'Pedigree Dogs', RSPCA, http://www.rspca.org.uk/adviceandwelfare /pets/dogs/health/pedigreedogs; 'Inherited Disorders in Cats', International Cat Care, http://www.icatcare.org:8080/advice/cat-breeds/inherited -disorders-cats.

16. For more on this theory, see my *Dog Sense: In Defence of Dogs* (New York: Basic Books, 2011).

17. Kit Yarrow, 'Millions on Pet Halloween Costumes? Why We Spend More and More on Pets', *Time*, 4 October 2012, http://business.time.com /2012/10/04/millions-on-pet-halloween-costumes-why-we-spend-more -and-more-on-pets.

18. Caitlin Moran, 'I'm Different When I'm Alone', *The Times*, 3 May 2014, http://www.thetimes.co.uk/tto/magazine/article4075131.ece. For the accusation that pets are child substitutes, see note 16.

19. Pets sell papers: see Matthew D. Atkinson, Maria Deam and Joseph E. Uscinski, 'What's a Dog Story Worth?', *Political Science*, October 2014, 1–5, doi: 10.1017/S1049096514001103.

20. Dr Ben Ambridge, 'Why Do We Believe in Homeopathy? Ten Tricks the Brain Plays on Us', *Telegraph*, 15 August 2014, http://www.telegraph .co.uk/culture/books/11012900/Why-do-we-believe-in-homeopathy-Ten -tricks-the-brain-plays-on-us.html.

21. Nina Shen Rastogi, 'The Trouble with Kibbles', *Slate*, 23 February 2010, http://www.slate.com/articles/health_and_science/the_green _lantern/2010/02/the_trouble_with_kibbles.html; Reuters, 'Killer Cats: Deadly Pets Murder Nearly 4 Billion Birds a Year', *Telegraph*, 30 January 2013, http://www.telegraph.co.uk/news/worldnews/northamerica/usa/9836471 /Killer-cats-deadly-pets-murder-nearly-4-billion-birds-a-year.html; 'Fish in Tanks? No, Thanks!', PETA, http://www.peta.org/issues/companion -animal-issues/companion-animals-factsheets/fish-tanks-thanks.

22. Not that we could blame the pets themselves, since, although we may cynically portray them (cats especially) as cocking a snook at our indulgences, comprehensive research into how their minds work has failed to discern any such capacity. Thus, even if they appeared to be exploiting our gullibility, in reality they could only be innocent participants.

23. Adam Sage, 'Cat-Loving Le Pen Shows Cuddly Side', *The Times*, 19 March 2016, 38.

24. One problem with modern intensive farming is that it seems to facilitate the removal of respect for the animal eaten.

25. 'Bird Feeding Proves Recession-Proof', British Trust for Ornithology, February 2012, http://www.bto.org/news-events/press-releases/bird -feeding-proves-recession-proof.

26. Associated Press, 'Fur Flies over 16th Century "Rocket Cats" Warfare Manual', *Guardian*, 6 March 2014, http://www.theguardian.com/books /2014/mar/06/fur-flies-rocket-cats-warfare-manual.

Chapter 1

1. The idea that our brains first became fully modern some 50,000 years ago rests on theories put forward by Professor Steven Mithen in *The*

Prehistory of the Mind: A Search for the Origins of Art, Religion and Science (London: Thames & Hudson, 1996).

2. For many more examples, see Chapter 4 of James Serpell, *In the Company of Animals* (Cambridge, UK: Canto, 1996); Francis Galton, 'The First Steps Towards the Domestication of Animals', *Transactions of the Ethnological Society* 3 (1865): 122–38; Peter Gray and Sharon Young, 'Human–Pet Dynamics in Cross-Cultural Perspective', *Anthrozoös* 24 (2011): 17–30.

3. Loretta Cormier, *Kinship with Monkeys: The Guajá Foragers of Eastern Amazonia* (New York: Columbia University Press, 2003).

4. Frederick Simoons and James Baldwin, 'Breast-Feeding of Animals by Women: Its Socio-cultural Context and Geographical Occurrence', *Anthropos* 77 (1982): 421–48. I am indebted to James Serpell for supplying me with this paper.

5. Laura Rival, 'The Growth of Family Trees: Understanding Huaorani Perceptions of the Forest', *Man*, n.s. 28 (1993): 635–52.

6. Lisa Maher et al., 'A Unique Human–Fox Burial from a Pre-Natufian Cemetery in the Levant (Jordan)', *PLoS ONE* 6 (2011): article e15815.

7. Philippe Erikson, 'The Social Significance of Pet Keeping Among Amazonian Indians', in *Companion Animals and Us: Exploring the Relationships Between People and Pets*, ed. Anthony Podberscek, Elizabeth Paul, and James Serpell (Cambridge: Cambridge University Press, 2000), 7–26.

8. Ibid.

9. James Gorman, 'The Big Search to Find Out Where Dogs Came From', *New York Times*, 18 January 2016, http://www.nytimes.com/2016/01/19/science/the-big-search-to-find-out-where-dogs-come-from.html.

10. See Chapter 7 in Darcy Morey, *Dogs: Domestication and the Development of a Social Bond* (Cambridge: Cambridge University Press, 2010).

11. Graduate student Angela Perri first proposed this theory at the 2010 Paris meeting of the International Conference of Archaeozoology. See Michael Balter, 'Burying Man's Best Friend, With Honour', *Science*, 17 September 2010, http://www.sciencemag.org/content/329/5998/1464.2.full.

12. For a more detailed account of the domestication of the cat, see my *Cat Sense* (New York: Basic Books, 2013).

13. Genesis 1:28 (New International Version).

14. James Serpell, '"Working Out the Beast": An Alternative History of Western Humaneness', in *Child Abuse, Domestic Violence and Animal Abuse: Linking the Circles of Compassion for Prevention and Intervention*, ed. Phil Arkow and Frank R. Ascione, Chapter 3 (West Lafayette, IN: Purdue University Press, 1999).

15. 'The Scholar and His Cat, Pangur Bán', translated from the Irish by Robin Flower, University of Pennsylvania, Department of Linguistics, https://www.ling.upenn.edu/~beatrice/pangur-ban.html.

16. Philip Stubbes, A *Christall Glasse for Christian Women* (London: John Wright, 1634).

17. Keith Thomas, *Man and the Natural World: Changing Attitudes in England, 1500–1800* (London: Allen Lane, 1983), 102.

18. Thomas Wright, *The Life of William Cowper* (New York: Haskell House Publishers, 1892).

19. Robert Hirst, ed. *Who Is Mark Twain?* (New York: Harper, 2009).

Chapter 2

1. Translated by Kathleen Kete from Alfred Barbou, *Le Chien: Son histoire, ses exploits, ses aventures* (Paris: Furne, Jouvet et Compagnie, 1883).

2. William Chambers, *The Story of a Long and Busy Life* (Edinburgh: W. & R. Chambers, 1882).

3. Royal Society for the Prevention of Cruelty to Animals, *Domestic Animals and Their Treatment* (London: RSPCA, 1857).

4. Susan Keaveney, 'Equines and Their Human Companions', *Journal of Business Research* 61 (2008): 444–54.

5. See, for example, Deborah Wells and Peter Hepper, 'Pet Ownership and Adults' Views on the use of Animals', *Society & Animals* 5 (1997): 45–63.

6. '*Cat Watch 2014*: The New *Horizon* Experiment', BBC, http://www .bbc.co.uk/programs/b04lcqvq; Anonymous, *The Power of Pets: A Summary of the Wide-Ranging Benefits of Companion Animal Ownership* (Artarmon, Australia: Australian Companion Animal Council, 1995).

7. Harold Herzog, 'Forty-Two Thousand and One Dalmatians: Fads, Social Contagion, and Dog Breed Popularity', *Society & Animals* 14 (2006): 383–97; Stefano Ghirlanda, Alberto Acerbi and Harold Herzog, 'Dog Movie Stars and Dog Breed Popularity: A Case Study in Media Influence on Choice', *PLoS ONE* 9 (2014): article e106565, doi: 10.1371/journal.pone .0106565.

8. See Ayaka Miura, John W. S. Bradshaw and Hajime Tanida, 'Attitudes Towards Assistance Dogs in Japan and the UK: A Comparison of College Students Studying Animal Care', *Anthrozoös* 15 (2002), 227–42; poster titled 'The Westernisation of Attitudes Towards Dogs in Japan', presented at the 2014 meeting of the International Society for Anthrozoology held in Vienna.

9. '2004 Pet Owner Survey', American Animal Hospital Association, https://www.aahanet.org/PublicDocuments/petownersurvey2004.pdf.

10. 'Genetic Savings and Clone Forced to Shut Down', *BioTechnology*, January 2009, http://biotechnology-industries.blogspot.co.uk/2009/01 /genetic-savings-and-clone-forced-to.html.

11. 'Animal Hoarding Case Study: Vikki Kittles', ALDF, 10 March 2012, http://aldf.org/resources/laws-cases/animal-hoarding-case-study-vikki

-kittles; Lindsey Erin Kroskob, 'Animal Hoarder Found Guilty', *Wyoming News*, 10 March 2012, http://www.wyomingnews.com/articles/2012/03/10 /news/01top_03-10-12.txt.

12. Gail Steketee et al., 'Characteristics and Antecedents of People Who Hoard Animals: An Exploratory Comparative Interview Study', *Review of General Psychology* 15 (2011): 114–24, doi: 10.1037/a0023484.

13. For the study in South Wales, see Jennifer Maher and Harriet Pierpoint, 'Friends, Status Symbols and Weapons: The Use of Dogs by Youth Groups and Youth Gangs', *Crime, Law and Social Change* 55 (2011): 405–20. For a robust account of the pit bull controversy in the United Kingdom, see Simon Hallsworth, 'Then They Came for the Dogs!', *Crime, Law and Social Change* 55 (2011): 391–403.

14. Jane Murray et al., 'Number and Ownership Profiles of Cats and Dogs in the UK', *Veterinary Record* 166 (2010): 163–8; Miho Nagasawa and Mitsuaki Ohta, 'The Influence of Dog Ownership in Childhood on the Sociality of Elderly Japanese Men', *Animal Science Journal* 81 (2010): 377–83.

15. James Serpell first demonstrated this connection in 'Childhood Pets and Their Influence on Adults' Attitudes', *Psychological Reports* 49 (1981): 651–4. It's been confirmed several times since, recently by Carri Westgarth et al., 'Family Pet Ownership During Childhood: Findings from a UK Birth Cohort and Implications for Public Health Research', *International Journal of Environmental Research and Public Health* 7 (2010): 3704–29, doi: 10.3390/ijerph7103704.

16. Kristen Jacobson et al., 'Genetic and Environmental Influences on Individual Differences in Frequency of Play with Pets Among Middle-Aged Men: A Behavioural Genetic Analysis', *Anthrozoös* 25 (2012): 441–56.

17. Ayaka Miura, John Bradshaw and Hajime Tanida. 'Childhood Experiences and Attitudes Towards Animal Issues: A Comparison of Young Adults in Japan and the UK', *Animal Welfare* 11 (2002): 437–48.

18. James Suzman, 'Sympathy for a Desert Dog', *Opinionator* (blog), *New York Times*, 31 August 2014, http://opinionator.blogs.nytimes.com //2014/08/31/sympathy-for-a-desert-dog.

Chapter 3

1. Karen Pallarito, '12 Ways Pets Improve Your Health', MSN.com, 15 August 2014, http://www.msn.com/en-us/health/wellness/12-ways-pets -improve-your-health/ss-AA2fsAf. See also Mandy Oaklander, 'Science Says Your Pet Is Good for Your Mental Health', *Time Health*, 6 April 2017, http://time.com/4728315/science-says-pet-good-for-mental-health.

2. Molly Crossman provides an excellent objective assessment of the (paucity of) science behind the psychological benefits of pets in 'Effects of

Interactions with Animals on Human Psychological Distress', *Journal of Clinical Psychology* (2016), doi:10.1002/jclp.22410.

3. Gerald Mallon, 'A Generous Spirit: The Work and Life of Boris Levinson', *Anthrozoös* 7 (1994): 224–31.

4. Erika Friedmann et al., 'Animal Companions and One-Year Survival of Patients After Discharge from a Coronary Care Unit', *Public Health Reports* 95 (1980): 307–12; Erika Friedmann and Sue Thomas, 'Pet Ownership, Social Support, and One-Year Survival After Acute Myocardial Infarction in the Cardiac Arrhythmia Suppression Trial (CAST)', *American Journal of Cardiology* 76 (1995): 1213–17; Gordon Parker et al., 'Survival Following an Acute Coronary Syndrome: A Pet Theory Put to the Test', *Acta Psychiatrica Scandinavica* 121 (2010): 65–70; Sarah Knight and Victoria Edwards, 'In the Company of Wolves: The Physical, Social, and Psychological Benefits of Dog Ownership', *Journal of Aging & Health* 20 (2008): 437–55; 'Pets and Health: Family Physician Survey', Human–Animal Bond Research Initiative, http://www.habri.org/2014-physician-survey.php.

5. For more information on therapy animals, see 'All About Therapy Animals', National Service Animal Registry, https://www.nsarco.com/therapy-animal-info.html.

6. Judith Simon Prager, PhD, 'Dolphin-Assisted Therapy: Something Magical in the Water', *Huffington Post*, 2011, http://www.huffingtonpost.com/judith-simon-prager-phd/dolphin-assisted-therapy_b_996389.html.

7. Lori Marino and Scott Lilienfeld, 'Dolphin-Assisted Therapy: More Flawed Data and More Flawed Conclusions', *Anthrozoös* 20 (2007): 239–49; Karsten Brensing, Katrin Linke and Dietmar Todt, 'Can Dolphins Heal by Ultrasound?', *Journal of Theoretical Biology* 225 (2003): 99–105.

8. 'The Case Against Dolphin Assisted Therapy', Whale and Dolphin Conservation, http://uk.whales.org/issues/case-against-dolphin-assisted-therapy; 'Dolphin Therapy and Autism', Research Autism, http://www.researchautism.net/interventions/64/dolphin-therapy-and-autism/Introduction.

9. Contrast Lissa Corcoran, 'Straight from the Horse's Mouth: Equine Assisted Psychotherapy', *Best Self Atlanta*, http://www.bestselfatlanta.com/sponsored-topic/life-enrichment/equine-assisted-psychotherapy/straight-from-the-horses-mouth-equine-assisted-psychotherapy.html, with Michael Anestis et al., 'Equine-Related Treatments for Mental Disorders Lack Empirical Support: A Systematic Review of Empirical Investigations', *Journal of Clinical Psychology* 70 (2014): 1115–32.

10. See, for example, a recent breakdown of the usefulness of animal-assisted interventions for post-traumatic stress disorder: Marguerite O'Haire, Noémie Guérin and Alison Kirkham, 'Animal-Assisted Intervention for Trauma: A Systematic Literature Review', *Frontiers in Psychology* 6 (August 2015), article 1121, doi: 10.3389/fpsyg.2015.01121, with a summary at

Marguerite E. O'Haire et al., 'Animal-Assisted Intervention for Trauma, Including Post-Traumatic Stress Disorder', HABRI Central, August 2015, https://habricentral.org/resources/48078.

11. Jennifer Limond, John Bradshaw and Magnus Cormack, 'Behaviour of Children with Learning Disabilities Interacting with a Therapy Dog', *Anthrozoös* 10 (1997): 84–9.

12. Marian Banks, Lisa Willoughby and William Banks, 'Animal-Assisted Therapy and Loneliness in Nursing Homes: Use of Robotic Versus Living Dogs', *Journal of the American Medical Directors Association* 9 (2008): 173–7; Hayley Robinson et al., 'The Psychosocial Effects of a Companion Robot: A Randomized Controlled Trial', *Journal of the American Medical Directors Association* 14 (2013): 661–7; Lori Marino, 'Construct Validity of Animal-Assisted Therapy and Activities: How Important Is the Animal in AAT?', *Anthrozoös* 25 (2012): S139–S151.

13. Hal Herzog, 'Does Animal-Assisted Therapy Really Work? What Clinical Trials Reveal About the Effectiveness of Four-Legged Therapists', *Psychology Today*, 17 November 2014, https://www.psychologytoday.com/blog/animals-and-us/201411/does-animal-assisted-therapy-really-work. Herzog refers to Janelle Nimer and Brad Lundahl, 'Animal-Assisted Therapy: A Meta-analysis', *Anthrozoös* 20 (2007): 225–38.

14. Giorgio Celani, 'Human Beings, Animals and Inanimate Objects: What Do People with Autism Like?', *Autism* 6 (2002): 93–102; Lillian Christon, Virginia Mackintosh and Barbara Myers, 'Use of Complementary and Alternative Medicine (CAM) Treatments by Parents of Children with Autism Spectrum Disorders', *Research in Autism Spectrum Disorders* 4 (2010): 249–59; Marguerite O'Haire, 'Animal-Assisted Intervention for Autism Spectrum Disorder: A Systematic Literature Review', *Journal of Autism and Developmental Disorders* 43 (2013): 1606–22.

15. Marguerite O'Haire et al., 'Social Behaviours Increase in Children with Autism in the Presence of Animals Compared to Toys', *PLoS ONE* 8, no. 2 (2013): e57010, doi: 10.1371/journal.pone.0057010; Marguerite O'Haire et al., 'Animals May Act as Social Buffers: Skin Conductance Arousal in Children with Autism Spectrum Disorder in a Social Context', *Developmental Psychobiology* (27 April 2015), doi: 10.1002/dev.21310.

16. Aimee Brannen, 'Unbreakable Bond Between an Autistic Little Girl and Her Cat', *Daily Mail*, 15 March 2016, http://www.dailymail.co.uk/femail/article-3491674/Autistic-six-year-old-paints-cycles-SWIMS-tabby-feline-helped-voice.html.

17. Sally Williams, 'The Cat That Saved My Autistic Son', *Sunday Times Magazine*, 8 February 2014, 26–32.

18. Marine Grandgeorge et al., 'Does Pet Arrival Trigger Prosocial Behaviours in Individuals with Autism?', *PLoS ONE* 7, no. 8 (2012): e41739, doi: 10.1371/journal.pone.0041739.

19. Val Elliott and Derek Milne, 'Patients' Best Friend?', *Nursing Times* 87, no. 6 (6 February 1991): 34–5.

20. Betty Carmack and Debra Fila, 'Animal-Assisted Therapy: A Nursing Intervention', *Nursing Management* 20, no. 5 (1989): 96–101.

21. Marian Banks and William Banks, 'The Effects of Animal-Assisted Therapy on Loneliness in an Elderly Population in Long-Term Care Facilities', *Journal of Gerontology: Medical Sciences* 57A (2002): M428–32.

22. Alessandra Berry et al., 'Developing Effective Animal-Assisted Intervention Programs Involving Visiting Dogs for Institutionalized Geriatric Patients: A Pilot Study', *Psychogeriatrics* 393 (2013): 143–50; Arline Siegel, 'Reaching the Severely Withdrawn Through Pet Therapy', *American Journal of Psychiatry* 118 (1962): 1045–46; Penny Bernstein, Erika Friedmann and Alessandro Malaspina, 'Animal-Assisted Therapy Enhances Resident Social Interaction and Initiation in Long-Term Care Facilities', *Anthrozoös* 13 (2000): 213–24.

23. Megan Souter and Michelle Miller, 'Do Animal-Assisted Activities Effectively Treat Depression? A Meta-analysis', *Anthrozoös* 20 (2007): 167–80.

24. Alison Cuff, 'Dementia: A Global Epidemic', *On Medicine*, 18 May 2015, http://blogs.biomedcentral.com/on-medicine/2015/05/18/dementia-global-epidemic.

25. Virginia Bernabei et al., 'Animal-Assisted Interventions for Elderly Patients Affected by Dementia or Psychiatric Disorders: A Review', *Journal of Psychiatric Research* 47 (2013): 762–73; Lena Nordgren and Gabriella Engström, 'Animal-Assisted Intervention in Dementia: Effects on Quality of Life', *Clinical Nursing Research* 23 (2014): 7–19.

26. Jitka Pikhartova, Ann Bowling and Christina Victor, 'Does Owning a Pet Protect Older People Against Loneliness?', *BMC Geriatrics* 14 (2014), http://www.biomedcentral.com/1471-2318/14/106.

27. For the effects of loneliness on health, see Julianne Holt-Lunstad et al., 'Loneliness and Social Isolation as Risk Factors for Mortality: A Meta-analytic Review', *Perspectives on Psychological Science* 10 (2015): 227–37.

28. Cheryl A. Krause-Parello, 'The Mediating Effect of Pet Attachment Support Between Loneliness and General Health in Older Females Living in the Community', *Journal of Community Health Nursing* 25 (2008): 1–14; Elsie Gulick and Cheryl Krause-Parello, 'Factors Related to Type of Companion Pet Owned by Older Women', *Journal of Psychosocial Nursing and Mental Health Services* 50 (2012): 30–37.

29. Ingela Enmarker et al., 'Depression in Older Cat and Dog Owners: The Nord-Trøndelag Health Study (HUNT)-3', *Aging and Mental Health* 19 (2014): 347–52; Roni Beth Tower and Maki Nokota, 'Pet Companionship and Depression: Results from a United States Internet Sample', *Anthrozoös* 19 (2006): 50–64; Krista Marie Clark Cline, 'Psychological Effects of Dog Ownership: Role Strain, Role Enhancement, and Depression',

Journal of Social Psychology 150 (2010): 117–31; Jasmin Peacock, Anna Chur-Hansen and Helen Winefield, 'Mental Health Implications of Human Attachment to Companion Animals', *Journal of Clinical Psychology* 68 (2012): 292–303.

30. Sharon Bolin, 'The Effect of Companion Animals During Conjugal Bereavement', *Anthrozoös* 1 (1987): 26–35; Thomas Garrity et al., 'Pet Ownership and Attachment as Supportive Factors in the Health of the Elderly', *Anthrozoös* 3 (1989): 35–44.

31. John Rogers, Lynette Hart and Ronald Boltz, 'The Role of Pet Dogs in Casual Conversations of Elderly Adults', *Journal of Social Psychology* 133 (1993): 263–77.

32. Susan Hunt, Lynette Hart and Richard Gomulkiewicz, 'Role of Small Animals in Interactions Between Strangers', *Journal of Social Psychology* 132 (1992): 245–56.

33. June McNicholas and Glyn Collis, 'Dogs as Catalysts for Social Interactions: Robustness of the Effect', *British Journal of Psychology* 91 (2000): 61–70; Deborah Wells, 'The Facilitation of Social Interactions by Domestic Dogs', *Anthrozoös* 17 (2004): 340–52.

34. Nicolas Guéguen and Serge Ciccotti, 'Domestic Dogs as Facilitators in Social Interaction: An Evaluation of Helping and Courtship Behaviours', *Anthrozoös* 21 (2008): 339–49; Sigal Tifferet et al., 'Dog Ownership Increases Attractiveness and Attenuates Perceptions of Short-Term Mating Strategy in Cad-like Men', *Journal of Evolutionary Psychology* 11 (2013): 121–9; Claire Budge et al., 'The Influence of Companion Animals on Owner Perception: Gender and Species Effects', *Anthrozoös* 9 (1996): 10–18.

35. See Lisa Wood, Billie Giles-Corti and Max Bulsara, 'The Pet Connection: Pets as a Conduit for Social Capital?', *Social Science & Medicine* 61 (2005): 1159–73.

36. Erika Friedman et al., 'Pet's Presence and Owner's Blood Pressure During Daily Lives of Owners with Pre- to Mild Hypertension', *Anthrozoös* 26 (2013): 535–50; Andrea Beetz et al., 'The Effect of a Real Dog, Toy Dog and Friendly Person on Insecurely Attached Children During a Stressful Task: An Exploratory Study', *Anthrozoös* 24 (2011): 349–68; see also Deborah Wells, 'The Effects of Animals on Human Health and Well-Being', *Journal of Social Issues* 65 (2009): 523–43.

37. See, for example, Warwick Anderson, Christopher Reid and Garry Jennings, 'Pet Ownership and Risk Factors for Cardiovascular Disease', *Medical Journal of Australia* 157 (1992): 298–301; James Serpell, 'Beneficial Effects of Pet Ownership on Some Aspects of Human Health and Behaviour', *Journal of the Royal Society of Medicine* 84 (1991): 717–20. Contrast these results with Gordon Parker et al., 'Survival Following an Acute Coronary Syndrome: A Pet Theory Put to the Test', *Acta Psychiatrica Scandinavica* 121 (2010): 65–70. Many such studies are based

on correlations, a statistical technique that is mathematically robust but in real life notoriously open to over-interpretation. For example, the web-site tylervigen.com shows a highly significant correlation between the age of Miss America and the number of murders by steam, hot vapours and hot objects in the same year, which is presumably just a fluke and not due to a few individuals becoming so disgusted by a 24-year-old beauty queen being preferred over her younger rivals that they decided to go on a kill-ing spree (see 'Age of Miss America Correlates with Murders by Steam, Hot Vapours, and Hot Objects', tylervigen.com, http://www.tylervigen.com /spurious-correlations). However, misleading correlations can also occur because of common underlying trends, such as the similarity between an-nual rates of suicide and spending on science research in the United States, both of which roughly doubled over the decade between 1999 and 2009, as the population grew and became more affluent. No correlation should ever be taken to imply cause and effect, however (im)plausible, although this is routinely done. The differences between owners and non-owners that may explain away the 'health benefits' are documented in Jessica Saunders, Layla Parast, Susan Babey and Jeremy Miles, 'Exploring the differences be-tween pet and non-pet owners: Implications for human–animal interaction research and policy', *PLoS ONE* 12(6) (2017): e0179494

38. See Hal Herzog, 'Why Do Human Friends (but Not Pets) Make Peo-ple Live Longer?', *Psychology Today*, July 2013, https://www.psychologytoday .com/blog/animals-and-us/201307/why-do-human-friends-not-pets-make -people-live-longer; Hal Herzog, 'Study Finds Dog Walkers Have More Bad Mental Health Days!', *Psychology Today*, February 2017, https://www .psychologytoday.com/blog/animals-and-us/201702/study-finds-dog-walkers -have-more-bad-mental-health-days.

39. Rebecca Utz, 'Walking the Dog: The Effect of Pet Ownership on Human Health and Health Behaviours', *Social Indicators Research* 116 (2014): 327–39; Glenn Levine et al., 'Pet Ownership and Cardiovascular Risk: A Scientific Statement from the American Heart Association', *Circulation* 127 (2013): 2353–63; Richard Gillum and Thomas Obisesan, 'Living with Companion Animals, Physical Activity and Mortality in a U.S. National Cohort', *Interna-tional Journal of Environmental Research and Public Health* 7 (2010): 2452–9; Carri Westgarth, Robert Christley and Hayley Christian, 'How Might We Increase Physical Activity Through Dog Walking? A Comprehensive Review of Dog Walking Correlates', *International Journal of Behavioural Nutrition and Physical Activity* 11 (2014): article 83, http://www.ijbnpa.org/content/11/1/83; Steven Moore et al., 'Leisure Time Physical Activity of Moderate to Vigorous Intensity and Mortality: A Large Pooled Cohort Analysis', *PLoS Medicine* 9, no. 11 (2012): e1001335, doi: 10.1371/journal.pmed.1001335.

40. See Hal Herzog's breakdown of the costs and economic benefits of pet keeping at 'Three Reasons Why Pets Don't Lower Health Care Costs',

Psychology Today, January 2016, https://www.psychologytoday.com/blog
/animals-and-us/201601/three-reasons-why-pets-dont-lower-health-care
-costs.

41. Francesca Solmi et al., 'Curiosity Killed the Cat: No Evidence of an
Association Between Cat Ownership and Psychotic Symptoms at Ages 13
and 18 Years in a UK General Population Cohort', *Psychological Medicine*
(2017), doi: 10.1017/S0033291717000125.

42. John McFadden, 'The Great Atopic Diseases Epidemic: Does Chemical Exposure Play a Role?', *British Journal of Dermatology* 166 (2015):
1156–7.

43. Andrew Fretzayas, Doxa Kotzia and Maria Moustaki, 'Controversial
Role of Pets in the Development of Atopy in Children', *World Journal of*
Pediatrics 9 (2013): 112–19.

44. Stanley Coren, 'Allergic Patients Do Not Comply with Doctors'
Advice to Stop Owning Pets', *British Journal of Medicine* 314 (15 February
1997): 517; Randi Bertelsen et al., 'Do Allergic Families Avoid Keeping
Furry Pets?', *Indoor Air* 20 (2010): 187–95.

Chapter 4

1. Recently academic attention has focused less on the fact that animals
can outnumber humans and more on the fact that male animals outnumber
female animals, starting right at the beginning with Beatrix Potter's characters. See Janice McCabe et al., 'Gender in Twentieth-Century Children's
Books: Patterns of Disparity in Titles and Central Characters', *Gender and*
Society 25, no. 2 (April 2011): 197–226; see also Carolyn Burke and Joby
Copenhaver, 'Animals as People in Children's Literature', *Language Arts*
81, no. 3 (2004): 205–13.

2. The quote is from Froma Walsh, 'Human–Animal Bonds II: The Role
of Pets in Family Systems and Family Therapy', *Family Process* 48, no. 4
(2009): 481–99. For another example, see Bill Strickland, 'The Benefits of
Pets: 5 Surprising Ways Pets Are Good for Kids – and Families Too!', *Par-*
ents, March 2008, http://www.parents.com/parenting/pets/kids/pets-good
-for-kids. For an example of the interaction between hygiene and asthma,
see Guang-hui Dong et al., 'Asthma and Asthma-Related Symptoms in
16,789 Chinese Children in Relation to Pet Keeping and Parental Atopy',
Journal of Investigative Allergology and Clinical Immunology 18, no. 3 (2008):
207–13.

3. See Gail Melson, 'Availability of and Involvement with Pets by Children: Determinants and Correlates', *Anthrozoös* 1 (1988): 45–52; Janine
Muldoon, Joanne Williams and Alistair Lawrence, '"Mum Cleaned It and I
Just Played with It": Children's Perceptions of Their Roles and Responsibili-
ties in the Care of Family Pets', *Childhood* 22 (2015): 201–6.

4. Marco Taubert and Burkhard Pleger, 'A Sound Children's Mind in a Healthy Children's Body', *Frontiers in Neuroscience* 8 (2014): article 143, doi: 10.3389/fnins.2014.00143; Ewa Miendlarzewska and Wiebke Trost, 'How Musical Training Affects Cognitive Development: Rhythm, Reward and Other Modulating Variables', *Frontiers in Neuroscience* 7 (2013), article 279, doi: 10.3389/fnins.2013.00279.

5. Colloquially, 'empathy' refers to a person's capacity to share and understand the feelings of others. When we see someone we are fond of crying, most of us spontaneously feel sad ourselves. If we witness the reuniting of a lost child and his mother, we may share a little of the relief and happiness we presume both must be feeling. But empathy is a complex emotion: some elements of it work at an almost intuitive, emotional level, while others require more thought, particularly the ability to identify and then imagine the changes in emotion that we observe in those around us. Studies of brain functioning back up this distinction: the more emotional parts of empathy are based in the mirror neurons, while cognitive empathy is linked to the 'theory of mind' systems, the part of the brain that enables us to realize that other people have minds and to construct ideas of what they might be thinking. The cognitive part enables us to interact with others in a sophisticated way: for example, we can empathize with a relative who has just been diagnosed with a life-threatening disease, even though we may have little knowledge of that particular condition and have never personally been in such a situation. We can also empathize (appreciate another's emotional state) without automatically sympathizing, perhaps if we feel that the other person's emotion is misplaced or inappropriate for the situation. However, in real life both emotional and cognitive empathy are usually activated simultaneously, so we may not notice that they're actually different. See Richard Koestner, Carol Franz and Joel Weinberger, 'The Family Origins of Empathetic Concern: A 26-Year Longitudinal Study', *Journal of Personality and Social Psychology* 58 (1990): 709–17; Mark Davis, Carol Luce, and Stephen Kraus, 'The Heritability of Characteristics Associated with Dispositional Empathy', *Journal of Personality* 62 (1994): 369–91. The persistence of autism in modern society has stimulated speculation that it may occasionally have been advantageous during our hunter-gatherer past, enabling individuals to forage more effectively when food was thinly spread, rendering group living a recipe for starvation. See Jared Reser, 'Conceptualizing the Autism Spectrum in Terms of Natural Selection and Behavioural Ecology: The Solitary Forager Hypothesis', *Evolutionary Psychology* 9 (2011): 207–38; University of Southern California, 'Autism May Have Had Advantages in Humans' Hunter-Gatherer Past, Researcher Believes', *Science Daily*, 3 June 2011, http://www.sciencedaily.com/releases/2011/06/110603122849.htm.

6. For an account of early American humane societies, see Monica Flegel, *Conceptualizing Cruelty to Children in Nineteenth-Century England:*

Literature, Representation, and the NSPCC (Farnham, UK: Ashgate Publishing, 2009), 61.

7. Sarah Knight et al., 'Human Rights, Animal Wrongs? Exploring Attitudes Towards Animal Use and Possibilities for Change', *Society & Animals* 18 (2010): 251–72; Tania Signal and Nicola Taylor, 'Attitude to Animals and Empathy: Comparing Animal Protection and General Community Samples', *Anthrozoös* 20 (2007): 125–30; Adelma Hills, 'The Motivational Bases of Attitudes Towards Animals', *Society & Animals* 1 (1993): 111–28.

8. See Malin Angantyr, Jakob Eklund, and Eric Hansen, 'A Comparison of Empathy for Humans and Empathy for Animals', *Anthrozoös* 24 (2011): 369–77.

9. Elizabeth Paul, 'Empathy with Animals and with Humans: Are They Linked?', *Anthrozoös* 13 (2000): 194–202.

10. 'Educator Resources', American Humane, https://www.american humane.org/fact-sheet/educator-resources.

11. Rebecca Purewal et al., 'Companion Animals and Child/Adolescent Development: A Systematic Review of the Evidence', *International Journal of Environmental Research and Public Health* 14 (2017): 234, doi:10.3390/ijerph14030234. For the early study, see Gail Melson, 'Child Development and the Human–Companion Animal Bond', *American Behavioural Scientist* 47 (2003): 31–9. For other reviews of child development, see Nienke Endenburg and Hein van Lith, 'The Influence of Animals on the Development of Children', *Veterinary Journal* 190 (2011): 208–14; Kelly Thompson and Eleonora Gullone, 'Promotion of Empathy and Prosocial Behaviour in Children Through Humane Education', *Australian Psychologist* 38 (2003): 175–82.

12. Robert Poresky, 'The Young Children's Empathy Measure: Reliability, Validity and Effects of Companion Animal Bonding', *Psychological Reports* 66 (1990): 931–6; 'Companion Animals and Other Factors Affecting Young Children's Development', *Anthrozoös* 9 (1996): 159–68; see also Beth Daly and Larry Morton, 'Children with Pets Do Not Show Higher Empathy: A Challenge to Current Views', *Anthrozoös* 16 (2003): 298–314; 'An Investigation of Human–Animal Interactions and Empathy as Related to Pet Preference, Ownership, Attachment, and Attitudes in Children', *Anthrozoös*, 19 (2006): 113–27; 'Empathic Differences in Adults as a Function of Childhood and Adult Pet Ownership and Pet Type', *Anthrozoös* 22 (2009): 371–82.

13. Brenda Bryant, 'The Neighbourhood Walk: Sources of Support in Middle Childhood', *Monographs of the Society for Research in Child Development* 50, no. 3 (1985).

14. Margaret Mead, 'Cultural Factors in the Cause and Prevention of Pathological Homicide', *Bulletin in the Menninger Clinic* 28 (1964): 11–22.

15. See Cathy Kangas, 'Animal Cruelty and Human Violence', *Huffington Post*, 18 January 2013, http://www.huffingtonpost.com/cathy-kangas/animal-cruelty-and-human-_b_2507551.html.

16. Arnold Arluke and Eric Madfis, 'Animal Abuse as a Warning Sign of School Massacres: A Critique and Refinement', *Homicide Studies* 18 (2014): 7–22.

17. Gary Fine, 'The Dirty Play of Little Boys', *Society* 24 (1986): 63–7.

18. 'The Link' appears to be a registered trademark of the American Humane Association; see 'The Link', American Humane Association, http://www.americanhumanesociety.net/interaction/professional-resources/the-link. See also Frank R. Ascione, *Children and Animals: Exploring the Roots of Kindness and Cruelty* (West Lafayette, IN: Purdue University Press, 2005). The pamphlet titled 'Understanding the Links: Child Abuse, Animal Abuse and Domestic Violence' appears not to be available from the NSPCC itself but is widely accessible on the Internet, including on the website of the Links Group (http://www.thelinksgroup.org.uk/wccms-resources/3/141bb3dc-e44c-11e4-9dd2-0050568626ea.pdf). Much of the same advice appears in 'Guidance on Investigating Child Abuse and Safeguarding Children', produced for the United Kingdom's Association of Chief Police Officers by the National Policing Improvement Agency in 2009, available on the CEOP website (https://www.ceop.police.uk/Documents/ACPOGuidance2009.pdf).

19. See Emily Patterson-Kane and Heather Piper, 'Animal Abuse as a Sentinel for Human Violence: A Critique', *Journal of Social Issues* 65 (2009): 589–614.

20. See, especially, Dorian Solot, 'Untangling the Animal Abuse Web', *Society & Animals* 5 (1997): 257–65, in which she also points out that virtually all research into animal abuse has focused on 'the link' rather than on the reasons why the animals were treated cruelly, which are likely to be far more complex than allowed for in the the link model. At the time of writing, electric shock collars are banned in Wales, Denmark, Norway, Sweden, Austria, Switzerland, Slovenia and Germany, and in some states in Australia; proposals are being considered for bans throughout the rest of the United Kingdom. See 'Electric Shock Collars', The Kennel Club, http://www.thekennelclub.org.uk/our-resources/kennel-club-campaigns/electric-shock-collars.

21. Paul Bracchi and Dennis Rice, 'I Want to Come Home Mummy', *Daily Mail*, 10 July 2009, http://www.dailymail.co.uk/femail/article-1198957/I-want-come-home-mummy-Aged-Jenny-torn-parents-social-workers-RSPCA-raid-Now-court-says-adopted-.html.

22. See, for example, Chiara Mariti et al., 'Improvement in Children's Humaneness Towards Nonhuman Animals Through a Project of Educational Anthrozoology', *Journal of Veterinary Behaviour: Clinical Applications*

and Research 6 (2011): 12–20; Christoph Randler, Eberhard Hummel and Pavol Prokop, 'Practical Work at School Reduces Disgust and Fear of Unpopular Animals', *Society & Animals* 20 (2012): 61–74.

23. See Bill Henry, 'Empathy, Home Environment and Attitudes Towards Animals in Relation to Animal Abuse', *Anthrozoös* 19 (2006): 17–34.

Chapter 5

1. Brand can be seen 'auctioning' Krusty at 'Russell Brand – Buy Love Here', video posted to YouTube by Russell Brand Stand Up Comedy, 26 May 2010, https://www.youtube.com/watch?v=cX7LtaUtCD8.

2. For a discussion of the legal issues, see Eithne Mills and Keith Akers, '"Who Gets the Cats . . . You or Me?" Analyzing Contact and Residence Issues Regarding Pets upon Divorce or Separation', *Family Law Quarterly* 36 (2002): 283–301.

3. The statistics on 'family' come from a 2006 survey of 3,000 US pet owners: 'Gauging Family Intimacy', Pew Research Center, 7 March 2006, http://www.pewsocialtrends.org/2006/03/07/gauging-family-intimacy. Thanks to the lessons learned during Katrina (see illustration), just a year later President George W. Bush signed into law H.R. 3858, the Pets Evacuation and Transportation Standards (PETS) Act. See William Brangham, 'Katrina 10 Years Later: How Did Katrina Change How We Evacuate Pets from Disaster?', *PBS Newshour*, 29 August 2015, http://www.pbs.org/newshour/rundown/hurricane-katrina-change-way-evacuate-pets-devastation.

4. The percentages are calculated from Nickie Charles and Charlotte Aull Davies, 'My Family and Other Animals: Pets as Kin', *Sociological Research Online* 13, no. 5 (2008) article 4, doi: 10.5153/sro.1798, adjusted using figures for pet ownership in Wales in 2008 taken from 'All about Pet Food', Pet Food Manufacturers' Association, http://www.pfma.org.uk/regional-pet-population.

5. Not all those people who depend on dogs to help them function regard them with affection. I once interviewed a blind person who, after some hesitation, eventually admitted that, while profoundly grateful to the guide dog organization – and his dog – for getting him mobile and independent once more, he had always been and still was phobic of dogs in general and dog saliva in particular. As a result, one part of his brain was forever focused on finding somewhere to wash his hands.

6. Charles and Davies, 'My Family and Other Animals'.

7. The press widely reported one study of cats as proving that they don't love their owners. In fact the experiment showed merely that many cats, unlike dogs, ignore their owners while they're exploring unfamiliar places. See John Bradshaw, 'Why Aren't Cats Loyal? You Asked Google – Here's

the Answer', *Guardian*, 30 September 2015, http://www.theguardian.com /commentisfree/2015/sep/30/why-arent-cats-loyal-google-answer.

8. 'Gauging Family Intimacy', Pew Research Center.

9. See Lawrence Kurdek, 'Pet Dogs as Attachment Figures for Adult Owners', *Journal of Family Psychology* 23 (2009): 439–46.

10. For more examples of how touch enhances the intimacy of owners' relationships with pets, see Alan Beck and Aaron Katcher, *Between Pets and People: The Importance of Animal Companionship* (West Lafayette, IN: Purdue University Press, 1996), 84–9.

11. See Chapter 6 of my *Dog Sense: In Defence of Dogs* (New York: Basic Books, 2011).

12. Sigal Zilcha-Mano, Mario Mikulincer, and Phillip Shaver, 'Pets as Safe Havens and Secure Bases: The Moderating Role of Pet Attachment Orientations', *Journal of Research in Personality* 46 (2012): 571–80.

13. The data come from Lawrence Kurdek, 'Pet Dogs as Attachment Figures for Adult Owners', *Journal of Family Psychology* 23 (2009): 439–46. Reassurance was assessed by asking owners to rate the following from one (strongly disagree) to seven (strongly agree): When I am feeling bad and need a boost, I turn to my dog to help me feel better; When I am disappointed, I turn to my dog to help me feel better; When something bad happens to me, I turn to my dog to help me feel better; When I am upset, I turn to my dog to help me feel better. Confidence was assessed from their ratings for the following: I can count on my dog to be there for me; I can depend on my dog to care about me no matter what; I can count on my dog's trustworthiness; I can count on my dog for comfort.

14. From Lilian Tzivian, Michael Friger and Talma Kushnir, 'Grief and Bereavement of Israeli Dog Owners: Exploring Short-Term Phases Pre- and Post-euthanization', *Death Studies* 38 (2014): 109–17.

15. For an example, see the talk I gave at the 2012 meeting of the International Society for Anthrozoology: 'John Bradshaw', video posted to YouTube by ISAZ2012, 14 August 2012, https://www.youtube.com/watch ?v=xLMb6X3Jab4.

16. Sheila Bonas, June McNicholas and Glyn Collis, 'Pets in the Network of Family Relationships: An Empirical Study', in *Companion Animals and Us: Exploring the Relationships Between People and Pets*, edited by Anthony Podberscek, Elizabeth Paul and James Serpell (Cambridge: Cambridge University Press, 2000), 209–36.

17. This quote and others in this and the following paragraphs come from Tzivian, Friger and Kushnir, 'Grief and Bereavement of Israeli Dog Owners', 109–17.

18. Apparently Pope Paul VI in the 1970s, not Pope Francis in 2014, uttered the words 'Paradise is open to all of God's creatures'; they went

unremarked at the time. See Madeleine Teahan, 'Pope Francis "Did Not Say Pets Go to Heaven"', *Catholic Herald*, 15 December 2014, http://www .catholicherald.co.uk/news/2014/12/15/pope-francis-did-not-say-pets-go -to-heaven. For examples from Hinduism, Buddhism and Shintoism, re- spectively, see Nivedita Kumar, 'Coping with the Death of a Pet', *Hindu*, 29 October 2014, http://www.thehindu.com/features/metroplus/coping -with-the-death-of-a-pet/article6545176.ece; biscuitsspace, 'Honouring Bis- cuit: A POWA Ceremony for Biscuit', Biscuit's Space, 29 August 2013, https://biscuitsspace.com/2013/08/29/honouring-biscuit-a-powa-ceremony -for-biscuit; 'Buddhist Pet Funerals', Quirky Japan Blog, https://qjphotos. wordpress.com/2010/05/17/buddhist-pet-funerals. The AIBO funeral is de- scribed by Miwa Suzuki, 'In Japan, Robot Dogs Are for Life – and Death', Phys.org, 25 February 2015, http://phys.org/news/2015-02-japan-robot-dogs -life-.html.

19. See Sherman Lee and Nicole Surething, 'Neuroticism and Religious Coping Uniquely Predict Distress Severity Among Bereaved Pet Owners', *Anthrozoös* 26 (2013): 61–76.

20. From Charles and Davies, 'My Family and Other Animals'.

21. Gretchen Reevy and Mikel Delgado, 'Are Emotionally Attached Companion Animal Caregivers Conscientious and Neurotic? Factors That Affect the Human–Companion Animal Relationship', *Journal of Applied Animal Welfare Science* 18 (2015): 239–58. For an overview of differences between men and women, see Hal Herzog, 'Gender Differences in Human– Animal Interactions: A Review', *Anthrozoös* 20 (2007): 7–21.

22. Allen McConnell et al., 'Friends with Benefits: On the Positive Consequences of Pet Ownership', *Journal of Personality and Social Psychol- ogy* 101 (2011): 1239–52; Sigal Zilcha-Mano, Mario Mikulincer and Phil- lip Shaver, 'An Attachment Perspective on Human–Pet Relationships: Conceptualization and Assessment of Pet Attachment Orientations', *Jour- nal of Research in Personality* 45 (2011): 345–57; Lisa Beck and Elizabeth Madresh, 'Romantic Partners and Four-Legged Friends: An Extension of Attachment Theory to Relationships with Pets', *Anthrozoös* 21 (2008): 43–56. John Archer and Jane Ireland also discuss attachment theory in 'The Development and Factor Structure of a Questionnaire Measure of the Strength of Attachment to Pet Dogs', *Anthrozoös* 24 (2011): 249–61.

23. Lisa Cavanaugh, Hillary Leonard and Debra Scammon, 'A Tail of Two Personalities: How Canine Companions Shape Relationships and Well-Being', *Journal of Business Research* 61 (2008): 469–79.

24. Charles and Davies, 'My Family and Other Animals'.

25. For an overview of the effects of problematic behaviour on owners' lives, see Victoria Voith, 'The Impact of Companion Animal Problems on Society and the Role of Veterinarians', *Veterinary Clinics of North America:*

Small Animal Practice 39 (2009): 327–45. For the survey referred to, see Carri Westgarth et al., 'Dog–Human and Dog–Dog Interactions of 260 Dog-Owning Households in a Community in Cheshire', *Veterinary Record* 162 (2008): 436–42.

26. Owners' ignorance or mishandling of the situation often exacerbates behavioural problems; see Andrew Jagoe and James Serpell, 'Owner Characteristics and Interactions and the Prevalence of Canine Behaviour Problems', *Applied Animal Behaviour Science* 47 (1996): 31–42.

27. From Leonard Simon, 'The Pet Trap: Negative Effects of Pet Ownership on Families and Individuals', in *The Pet Connection: Its Influence on Our Health and Quality of Life*, ed. Robert Anderson, Benjamin Hart and Lynette Hart (Minneapolis: Center to Study Human–Animal Relationships and Environments, University of Minnesota, 1984), 226–40.

28. It seems that the likelihood of a child's death leading to the parents' separation has been overplayed somewhat. See Catherine Rogers et al., 'Long-Term Effects of the Death of a Child on Parents' Adjustment in Midlife', *Journal of Family Psychology* 22 (2008): 203–11.

29. All quotes come from Susan Keaveney, 'Equines and Their Human Companions', *Journal of Business Research* 61 (2008): 444–54.

30. The grief that follows the death of a human family member is maladaptive – for example, it compromises immune function – and so is difficult to explain in evolutionary terms. Psychologists maintain that it is an unavoidable, if extreme, side effect of the anxiety we feel when temporarily separated from a loved one. See John Archer, *The Nature of Grief* (London: Routledge, 1999).

31. When I entered 'pet family' into a well-known search engine, the first 100 images (excluding logos) consisted of the following: 37 pictures of two- or three-generational families and their pet or pets; 30 of one or more adults with a selection of pets, including a scuba-diving dog and a capybara (three of the humans appeared to be naked, with their pets positioned strategically); 29 of children with 'their' pets (one was of a baby wrapped in a boa constrictor); 2 of idealized multispecies families (dogs with cats/rabbits/parrots); and only 2 of single-species families – of dogs.

32. For a sociologist's perspective, see Nickie Charles, 'Post-human families? Dog–Human Relations in the Domestic Sphere', *Sociological Research Online* 21 (3), 8 (31 August 2016), http://www.socresonline.org.uk/21/3/8 .html.

Chapter 6

1. 'PDSA Order of Merit', PDSA, https://www.pdsa.org.uk/what-we-do /animal-honours/the-order-of-merit; Anna Swartz, 'Brave Cat Wins Award

for Most Heroic . . . Dog', The Dodo, 22 June 2015, https://www.thedodo
.com/cat-wins-hero-dog-award-after-saving-her-boys-life-1211441676.html.

2. Richard Alleyne, 'The Legend of Greyfriars Bobby Really Is a
Myth', Telegraph, 3 August 2011, http://www.telegraph.co.uk/news/uknews
/8679341/The-legend-of-Greyfriars-Bobby-really-is-a-myth.html.

3. Fritz Heider and Marianne Simmel, 'An Experimental Study of Apparent Behaviour', American Journal of Psychology 57 (1944): 243–59.

4. For more details, see the following papers : Thalia Wheatley, Shawn
Milleville and Alex Martin, 'Understanding Animate Agents: Distinct
Roles for the Social Network and Mirror System', Psychological Science
18 (2007): 469–74; Brian Scholl and Patrice Tremoulet, 'Perceptual
Causality and Animacy', Trends in Cognitive Sciences 4 (2000): 299–309;
Andrea Heberlein and Ralph Adolphs, 'Impaired Spontaneous Anthro-
pomorphizing Despite Intact Perception and Social Knowledge', Proceed-
ings of the National Academy of Sciences of the United States of America 101
(2004): 7487–91; Naoyuki Osaka, Takashi Ikeda and Mariko Osaka,
'Effect of Intentional Bias on Agency Attribution of Animated Motion:
An Event-Related fMRI Study', PLoS ONE 7 (2012): e49053.

5. Adapted from Maya Mathur and David Reichling, 'Navigating a Social World with Robot Partners: A Quantitative Cartography of the Un-
canny Valley', Cognition 146 (2016): 22–32. Faces are not the only feature
of robots that trigger anthropomorphism: what they say may be more important; see Sara Kiesler and Jennifer Goetz, 'Mental Models and Cooperation with Robotic Assistants', Carnegie Mellon University, http://www
.cs.cmu.edu/~flo/papers/robot_chi_nonanon.pdf.

6. Linda Lyons, 'One-Third of Americans Believe Dearly May Not
Have Departed', Gallup, 12 July 2005, http://www.gallup.com/poll/17275
/onethird-americans-believe-dearly-may-departed.aspx; 'Ghosts & UFOs',
Association for the Scientific Study of Anomalous Phenomena, 2013, www
.assap.ac.uk/newsite/Docs/Ghost%20UFO%20Survey%202013.pdf; '18%
of Brits Believe in Possession by the Devil', YouGov, https://yougov.co.uk
/news/2013/09/27/18-brits-believe-possession-devil-and-half-america.

7. See 'A Third of British Adults Don't Believe in a Higher Power',
YouGov, 12 February 2015, https://yougov.co.uk/news/2015/02/12/third
-british-adults-dont-believe-higher-power.

8. For an evolutionary exploration of shamanism, see Bernard Crespi
and Kyle Summers, 'Inclusive Fitness Theory for the Evolution of Religion', Animal Behaviour 92 (2014): 313–23.

9. For more on the evolutionary approach to religion, see Elizabeth Cu-
lotta, 'On the Origin of Religion', Science (Washington) 326 (6 November
2009): 784–7.

10. See Stewart Guthrie, Faces in the Clouds: A New Theory of Religion (New York: Oxford University Press, 1993); Pascal Boyer, Religion

Explained: The Evolutionary Origins of Religious Thought (New York: Basic Books, 2001).

11. See Justin Barrett and Jonathan Lanman, 'The Science of Religious Beliefs', *Religion* 38 (2008): 109–24; Lee Kirkpatrick, 'The Role of Evolutionary Psychology Within an Interdisciplinary Science of Religion', *Religion* 41 (2011): 329–39.

12. From Justin Barrett and Frank Keil, 'Conceptualizing a Nonnatural Entity: Anthropomorphism in God Concepts', *Cognitive Psychology* 31 (1996): 219–47. The quote from Charles Darwin comes from *The Descent of Man, and Selection in Relation to Sex*, 2nd edn (London: John Murray, 1874), 65.

13. The researchers preferred London to the United States for this study because of its wide ethnic and religious mix of children; see Deborah Kelemen and Cara DiYanni, 'Intuitions About Origins: Purpose and Intelligent Design in Children's Reasoning About Nature', Journal of *Cognition and Development* 6 (2005): 3–31; see also Deborah Kelemen, 'Are Children "Intuitive Theists"? Reasoning About Purpose and Design in Nature', *Psychological Science* 15 (2004): 295–301.

14. In functional magnetic resonance imaging (fMRI), subjects lie in a brain scanner, which detects which parts of their brains are receiving the most oxygen – and therefore, by inference, are the most active. The study that asked subjects to think about God was Dimitrios Kapogiannis et al., 'Cognitive and Neural Foundations of Religious Belief', *Proceedings of the National Academy of Sciences of the United States of America* 106 (2009): 4876–81.

15. For an example, see Julia Hur, Minjung Koo and Wilhelm Hofmann, 'When Temptations Come Alive: How Anthropomorphism Undermines Self-Control', *Journal of Consumer Research* 42 (2015): 340–58. The M&Ms quote comes from 'M&M's Characters', Fandom, http://chocolate .wikia.com/wiki/M&M's_Characters.

16. For an experiment showing that even adults are instinctively egocentric, see Nicholas Epley, Carey Morewedge and Boaz Keysar, 'Perspective Taking in Children and Adults: Equivalent Egocentrism but Differential Correction', *Journal of Experimental Social Psychology* 40 (2004): 760–68.

17. The theoretical approach to anthropomorphism adopted here owes much to the writings of Adam Waytz of Harvard University and Nicholas Epley at Chicago. For reviews, see Nicholas Epley, Adam Waytz and John T. Cacioppo, 'On Seeing Human: A Three-Factor Theory of Anthropomorphism', *Psychological Review* 114 (2007): 864–86; Adam Waytz et al., 'Causes and Consequences of Mind Perception', *Trends in Cognitive Sciences* 14 (2010): 383–8.

18. For the details of the ball-bearing study, see Justin Barrett and Amanda Hankes Johnson, 'The Role of Control in Attributing Intentional

Agency to Inanimate Objects', *Journal of Cognition & Culture* 3 (2004): 208–17. For the fMRI study, see Adam Waytz et al., 'Making Sense by Making Sentient: Effectance Motivation Increases Anthropomorphism', *Journal of Personality and Social Psychology* 99 (2010): 410–35.

19. See Nicholas Epley et al., 'Creating Social Connection Through Inferential Reproduction: Loneliness and Perceived Agency in Gadgets, Gods, and Greyhounds', *Psychological Science* 19 (2008): 114–20; see also Nikolina Duvall Antonacopoulos and Timothy Pychyl, 'The Possible Role of Companion-Animal Anthropomorphism and Social Support in the Physical and Psychological Health of Dog Guardians', *Society & Animals* 18 (2010): 379–95; Elizabeth Paul et al., 'Sociality Motivation and Anthropomorphic Thinking About Pets', *Anthrozoös* 27 (2014): 499–512. A recent study was unable to replicate Epley's: Gillian M. Sandstrom and Elizabeth W. Dunn, 'Replication of "Creating Social Connection Through Inferential Reproduction: Loneliness and Perceived Agency in Gadgets, Gods, and Greyhounds" by Nick Epley, Scott Akalis, Adam Waytz, and John T. Cacioppo (2008, *Psychological Science*)', Open Science Framework, https://osf.io/m5a2c.

20. Harriet Cullen et al., 'Individual Differences in Anthropomorphic Attributions and Human Brain Structure', *Social Cognitive and Affective Neuroscience* 9 (2014): 1276–80.

21. For details of the experiments that underlie these findings, see Adam Waytz, John Cacioppo and Nicholas Epley, 'Who Sees Human? The Stability and Importance of Individual Differences in Anthropomorphism', *Perspectives on Psychological Science* 5 (2010): 219–32; Kim-Pong Tam, Sau-Lai Lee and Melody Manchi Chao, 'Saving Mr. Nature: Anthropomorphism Enhances Connectedness to and Protectiveness Towards Nature', *Journal of Experimental Social Psychology* 49 (2013): 514–21.

22. Christina Brown and Julia McLean, 'Anthropomorphizing Dogs: Projecting One's Own Personality and Consequences for Animal Rights', *Anthrozoös* 28 (2015): 73–86; see also Virginia Kwan, Samuel Gosling and Oliver John, 'Anthropomorphism as a Special Case of Social Perception: A Cross-Species Social Relations Model Analysis of Humans and Dogs', *Social Cognition* 26 (2008): 129–42.

23. The quote from Xenophanes comes from 'Clement of Alexandria, The Stromata, or Miscellanies, Book V', Early Christian Writings, http://www.earlychristianwritings.com/text/clement-stromata-book5.html. For a history of animated film, see Frank Thomas and Ollie Johnson, *The Illusion of Life: Disney Animation* (New York: Hyperion, 1981). Kathleen Gerbas et al. analyse the Furry phenomenon in 'Furries from A to Z: (Anthropomorphism to Zoomorphism)', *Society & Animals* 16 (2008): 197–222.

24. From Edward Payson Evans, *The Criminal Prosecution and Capital Punishment of Animals* (New York: E. P. Dutton, 1906), 144–5.

25. See Beryl Rowland, *Animals with Human Faces* (Knoxville: University of Tennessee Press, 1973).

26. Robert Mitchell, Nicholas Thompson and Lyn Miles, eds., *Anthropomorphism, Anecdotes and Animals* (Albany: State University of New York Press, 1997), lays out the debate over the scientific legitimacy of anthropomorphic language.

27. Nancy Spears, John Mowen and Goutam Chakraborty, 'Symbolic Role of Animals in Print Advertising: Content Analysis and Conceptual Development', *Journal of Business Research* 37 (1996): 87–95; Nancy Spears and Richard Germain, 'The Shifting Role and Face of Animals in Print Advertisements in the Twentieth Century', *Journal of Advertising* 36 (2007): 19–33. The quotation comes from Judith Kerr, *The Tiger Who Came to Tea* (London: William Collins Sons & Co., 1968).

28. Scott Plous, 'Psychological Mechanisms in the Human Use of Animals', *Journal of Social Issues* 49 (1993): 11–52.

29. See Valerie Sims et al., 'Eye Movements When Judging Affect in Cats and Dogs' (abstract, International Society for Anthrozoology 14th Annual Conference, Niagara Falls, 2005), 54–5; Sadahiko Nakajima, 'Dogs and Owners Resemble Each Other in the Eye Region', *Anthrozoös* 26 (2013): 551–6; see also Julie Hecht and Alexandra Horowitz, 'Seeing Dogs: Human Preferences for Dog Physical Attributes', *Anthrozoös* 28 (2015): 153–63; Christine Looser and Thalia Wheatley, 'The Tipping Point of Animacy: How, When, and Where We Perceive Life in a Face', *Psychological Science* 21 (2010): 1854–62. For 'eyebrow raising' in dogs, see Bridget Waller et al., 'Paedomorphic Facial Expressions Give Dogs a Selective Advantage', *PLoS ONE* 8 (2013), article e82686. They interpreted the eyebrow raise as enhancing the dog's 'baby-faced' appeal (see Chapter 7), but since asymmetries reduce the attractiveness of babies' faces, this is more likely an example of pure anthropomorphic appeal.

30. From Daniel Chamovitz, *What a Plant Knows: A Field Guide to the Senses of Your Garden – and Beyond* (New York: Scientific American, 2012), 43–7. For cultural differences between 'folk biologies', see Scott Atran, Douglas Medin and Norbert Ross, 'Evolution and Devolution of Knowledge: A Tale of Two Biologies', *Journal of the Royal Anthropological Institute* 10 (2004): 395–420.

31. The 'blob' study and the nearby illustration come from Carey Morewedge, Jesse Preston and Daniel Wegner, 'Timescale Bias in the Attribution of Mind', *Journal of Personality and Social Psychology* 93 (2007): 1–11. For species differences, see also Timothy Eddy, Gordon Gallup, and Daniel Povinelli, 'Attribution of Cognitive States to Animals: Anthropomorphism in Comparative Perspective', *Journal of Social Issues* 49 (1993): 87–101

32. Interestingly, although in English the word 'emotion' is used for all emotions, from the simplest to the most complex, some other languages

make a greater distinction. In French, 'emotion' is used for (vertebrate) 'basic' emotions, while 'sentiment' is reserved for emotions believed to be uniquely human: a similar distinction can be found in other Romance languages, such as Spanish. Hungarian is a Uralic language (i.e., not Romance).

33. Data come from Paul Morris, Christine Doe and Emma Godsell, 'Secondary Emotions in Non-primate Species? Behavioural Reports and Subjective Claims by Animal Owners', *Cognition & Emotion* 22 (2008): 3–20; Veronika Konok, Krisztina Nagy, and Ádám Miklósi, 'How Do Humans Represent the Emotions of Dogs? The Resemblance Between the Human Representation of the Canine and the Human Affective Space', *Applied Animal Behaviour Science* 162 (2015): 37–46.

34. Paul Morris, Sarah Knight and Sarah Lesley, 'Belief in Animal Mind: Does Familiarity with Animals Influence Beliefs About Animal Emotions?', *Society & Animals* 20 (2012): 211–24; Paul Morris, Margaret Fidler and Alan Costall, 'Beyond Anecdotes: An Empirical Study of "Anthropomorphism"', *Society & Animals* 8 (2000): 151–65.

35. I described this study at the 2012 conference of the International Society for Anthrozoology in Cambridge; see 'John Bradshaw', video posted to YouTube by ISAZ2012, 14 August 2012, https://www.youtube.com/watch ?v=xLMb6X3Jab4.

36. See Rupert Sheldrake, *Dogs That Know When Their Owners Are Coming Home: And Other Unexplained Powers of Animals* (London: Hutchinson, 1999). For a scientific perspective, see Richard Wiseman, Matthew Smith and Julie Milton, 'Can Animals Detect When Their Owners Are Returning Home? An Experimental Test of the "Psychic Pet" Phenomenon', *British Journal of Psychology* 89 (1998): 453–62, http://www.richardwiseman.com /resources/petsBJP.pdf.

37. Alexandra Horowitz describes her experiment in *Inside of a Dog* (New York: Scribner, 2009); for more on the 'dominance myth', see my *Dog Sense: In Defence of Dogs* (New York: Basic Books, 2011) and my *Psychology Today* blog, *Pets and Their People* (http://www.psychologytoday.com/blog /pets-and-their-people).

Chapter 7

1. For the effects of cuteness on mothers, see Judith Langlois et al., 'Infant Attractiveness Predicts Maternal Behaviours and Attitudes', *Developmental Psychology* 31 (1995): 464–72.

2. The list comes from Irenäus Eibl-Eibesfelt, *Ethology: The Biology of Behaviour*, trans. Erich Klinghammer (New York: Holt Reinhart & Winston, 1970), 431–2, which summarizes Lorenz's original 1943 paper (also in German). For the cartoon study, see Sarah Sternglanz, James Gray and

Melvin Murakimi, 'Adult Preferences for Infantile Facial Features: An Ethological Approach', *Animal Behaviour* 25 (1977): 108–15.

3. Christine Parsons et al., 'The Motivational Salience of Infant Faces Is Similar for Men and Women', *PLoS ONE* 6 (2011): e20632.

4. Jessika Golle et al., 'Preference for Cute Infants Does Not Depend on Their Ethnicity or Species: Evidence from Hypothetical Adoption and Donation Paradigms', *PLoS ONE* 10 (2015), article e0121554.

5. For a review of the effects of *Kindchenschema* on human behaviour and the human brain, see Lizhu Luo et al., 'Neural Systems and Hormones Mediating Attraction to Infant and Child Faces', *Frontiers in Psychology* 6 (2015), article 970.

6. ACC: anterior cingulate cortex; AMY: amygdala; FFG: fusiform gyrus; GP: Globus pallidus; INS: insula; IPS: intraparietal sulcus; MCC: medial cingulate cortex; NAcc: nucleus accumbens; OFC: orbitofrontal cortex; PAG: periaqueductal grey; PCU: precuneus; SFG: superior frontal gyrus; SMA: supplementary motor area; SNi: substantia nigra; THA: thalamus.

7. Oriana Aragón et al., 'Dimorphous Expressions of Positive Emotion: Displays of Both Care and Aggression in Response to Cute Stimuli', *Psychological Science* 26 (2015): 259–73.

8. See Bonniebrook Gallery, Museum and Homestead website (http://www.roseoneill.org/mainpage.html).

9. See Stephen Jay Gould, 'A Biological Homage to Mickey Mouse', in *The Panda's Thumb: More Reflections in Natural History* (New York: Norton, 1980).

10. Leo Lewis, 'Kitty Chases the Mouse into the Middle Kingdom', *The Times*, 10 May 2011, 41. The Japanese version of cuteness, *kawaii*, pervades much of Japanese culture and encompasses more than simple infantile features. For example, magazines and comics adopted a decorative style of handwriting first devised by teenage girls, sometimes referred to as 'kitten writing'.

11. In the nineteenth century, toy bears were almost lifelike, but in 1902 (according to manufacturer Steiff GmbH) the first poseable, plush-covered bears with jointed arms appeared in Germany. In November of that same year, on the other side of the Atlantic, a cartoon appeared in the *Washington Post*, depicting President Theodore 'Teddy' Roosevelt refusing to shoot a black bear. As the story goes, the governor of Mississippi had invited him on a bear hunt, but when no wild bear could be found, his assistants procured a bear and tied it to a tree. Roosevelt refused to shoot it on the grounds that this would be unsportsmanlike. An enterprising toymaker in Brooklyn then obtained the president's permission to name his line of toy bears 'Teddy's Bear'. The name stuck, and by 1906 both the domestic product and the imported Steiff bears had become known generically as

teddy bears. See 'Steiff: The Story', Steiff, http://www.steiffusa.com/steiff
-the-story; 'The Story of the Teddy Bear', National Park Service, http://
www.nps.gov/thrb/learn/historyculture/storyofteddybear.htm.

12. One of the researchers being a distinguished Fellow of the Royal
Society, Professor Robert Hinde CBE. For a description of the teddy-bear
study, see R. A. Hinde and L. A. Barden, 'The Evolution of the Teddy
Bear', *Animal Behaviour* 33 (1985): 1372–3.

13. Paul Morris, Vasu Reddy and R. C. Bunting, 'The Survival of the
Cutest: Who's Responsible for the Evolution of the Teddy Bear?', *Animal
Behaviour* 50 (1995): 1697–1700.

14. See Angelo Gazzano et al., 'Dogs' Features Strongly Affect People's
Feelings and Behaviour Towards Them', *Journal of Veterinary Behaviour* 8
(2013): 213–20; Julie Hecht and Alexandra Horowitz, 'Seeing Dogs: Human Preferences for Dog Physical Attributes', *Anthrozoös* 28 (2015): 153–
63; Anthony Little, 'Manipulation of Infant-like Traits Affects Perceived
Cuteness of Infant, Adult and Cat Faces', *Ethology* 118 (2012): 775–82.

15. Paul Spindler, 'Studien zur Vererbung von Verhaltensweisen: 3. Ver-
halten gegenüber jungen Katzen', *Anthropologischer Anzeiger* 25 (1961): 60–
80, kindly translated into English for me by Dr Barbara Schöning.

16. Gary Sherman, Jonathan Haidt and James Coan, 'Viewing Cute Im-
ages Increases Behavioural Carefulness', *Emotion* 9 (2009): 282–6; see also
Hiroshi Nittono et al., 'The Power of Kawaii: Viewing Cute Images Promotes a Careful Behaviour and Narrows Attentional Focus', *PLoS ONE* 7
(2012), article e46362. This seems to be a strong and robust effect, which
I was able to replicate on TV for the series *CatWatch 2014*, aka *The Truth
About Cats*, using a 'Don't Buzz The Wire' game in place of 'Operation'.
See 'Do Kittens Help You Focus?', Nat Geo Wild, http://channel.national-
geographic.com/wild/the-truth-about-cats/videos/do-kittens-help-you-focus.

17. For a discussion of the possible role of play in the domestication of
the dog, see Nicola Rooney and John Bradshaw, 'Canine Welfare Science:
An Antidote to Sentiment and Myth', in *Domestic Dog Cognition and Behaviour*, ed. Alexandra Horowitz (Berlin, Springer-Verlag, 2014), 261.

18. Gergana Nenkov and Maura Scott, '"So Cute I Could Eat It Up":
Priming Effects of Cute Products on Indulgent Consumption', *Journal of
Consumer Research* 41 (2014): 326–41.

19. For the study referred to, see Rowena Packer et al., 'Impact of Facial Conformation on Canine Health: Brachycephalic Obstructive Airway
Syndrome', *PLoS ONE* 10 (2015), article e0137496; see also James Serpell, 'Anthropomorphism and Anthropomorphic Selection – Beyond the
"Cute Response"', *Society & Animals* 11 (2003): 83–100. For the congenital
conditions suffered by pugs, see 'Genetic Welfare Problems of Companion
Animals', Universities Federation for Animal Welfare, http://www.ufaw
.org.uk/dogs/pug.

20. Pinar Thorn et al., 'The Canine Cuteness Effect: Owner-Perceived Cuteness as a Predictor of Human–Dog Relationship Quality', *Anthrozoös* 28 (2015): 569–85.

21. Little, 'Manipulation of Infant-like Traits Affects Perceived Cuteness of Infant, Adult and Cat Faces'; Jessika Golle et al., 'Sweet Puppies and Cute Babies: Perceptual Adaptation to Babyfacedness Transfers Across Species', *PLoS ONE* 8 (2013), article e58248.

22. Tobias Brosch, David Sander and Klaus Scherer, 'That Baby Caught My Eye . . . Attention Capture by Infant Faces', *Emotion* 7 (2007): 685–9; Vincenzo Senese et al., 'Human Infant Faces Provoke Implicit Positive Affective Responses in Parents and Non-parents Alike', *PLoS ONE* 8 (2013), article e80379; Andrea Caria et al., 'Species-Specific Response to Human Infant Faces in the Premotor Cortex', *NeuroImage* 60 (2012): 884–93.

23. Luke Stoeckel et al., 'Patterns of Brain Activation When Mothers View Their Own Child and Dog: An fMRI Study', *PLoS ONE* 9 (2014), article e107205.

24. Although baby talk is rather consistent across cultures, there are variations – for example, not all employ the high-pitched voice. The description given is for motherese in American English, since most studies of doggerel to date have examined native speakers of that language. Some suggest that the term 'motherese' may be sexist and we should use more gender-neutral terms like 'child-directed language', but since mothers appear to have invented it, I see no harm in giving credit where credit is due!

25. See Robert Mitchell, 'Americans' Talk to Dogs: Similarities and Differences with Talk to Infants', *Research on Language and Social Interaction* 34 (2001): 183–210; Kathy Hirsh-Pasek and Rebecca Treiman, 'Doggerel: Motherese in a New Context', *Journal of Child Language* 9 (1982): 229–37; Naoko Koda, 'Anthropomorphism in Japanese Women's Status Terms Used in Talk to Potential Guide Dogs', *Anthrozoös* 14 (2001): 109–11. One study done in Italy has confirmed the use of 'doggerel' by both women and, to a lesser extent, men in that language: Emanuela Prato-Previde, Gaia Fallani and Paola Valsecchi, 'Gender Differences in Owners Interacting with Pet Dogs: An Observational Study', *Ethology* 112 (2006): 64–73.

26. Denis Burnham, Christine Kitamura, and Uté Vollmer-Conna, 'What's New, Pussycat? On Talking to Babies and Animals', *Science* (Washington) 296 (2002): 1435; Valerie Sims and Matthew Chin, 'Responsiveness and Predicted Intelligence as Predictors of Speech Addressed to Cats', *Anthrozoös* 15 (2002): 166–77.

27. Nan Xu et al., 'Vowel Hyperarticulation in Parrot-, Dog- and Infant-Directed Speech', *Anthrozoös* 26 (2013): 373–80.

28. Aaron Pepe et al., 'Go, Dog, Go: Maze Training AIBO vs. a Live Dog: An Exploratory Study', *Anthrozoös* 21 (2008): 71–83.

Chapter 8

1. Phi Tran, 'Average Life of US Mobile Phone Is 18 Months', Adweek, 25 March 2013, http://www.adweek.com/socialtimes/33775/185697.

2. 'Man's Worst Friend: Average Dog Causes 2,000 Family Arguments in Its Lifetime', Daily Mail, 12 January 2012, http://www.dailymail.co .uk/news/article-2084835/Mans-worst-friend-Average-dog-causes-2-000 -family-arguments-lifetime.html. Children cause more arguments than pets do, on average.

3. Victoria Voith, 'The Impact of Companion Animal Problems on Society and the Role of Veterinarians', Veterinary Clinics of North America: Small Animal Practice 39 (2009): 327–45.

4. From Robert Hanks, 'The Tyranny of a Dog's Turds', Guardian, 23 March 2012, http://www.theguardian.com/commentisfree/2012/mar/23 /tyranny-of-dogs-turds.

5. The quotation comes from Nickie Charles and Charlotte Aull Davies, 'My Family and Other Animals: Pets as Kin', Sociological Research Online 13, no. 5 (2008), article 4, doi: 10.5153/sro.1798. For the survey about unwanted behaviour among cats, see John Bradshaw, Rachel Casey and Johanna MacDonald, 'The Occurrence of Unwanted Behaviour in the Pet Cat Population', Proceedings of the Companion Animal Behaviour Therapy Study Group (2000): 41–2.

6. Anthony Podberscek, 'Positive and Negative Aspects of Our Relationship with Companion Animals', Veterinary Research Communications 30 (Suppl. 1) (2006): 21–7; 'Provisional Monthly Topic of Interest: Admissions Caused by Dogs and Other Mammals', Health and Social Care Information Centre, May 2015, http://www.hscic.gov.uk/catalogue/PUB 17615/prov-mont-hes-admi-outp-ae-April%202014%20to%20February %202015-toi-rep.pdf.

7. 'Nonfatal Fall-Related Injuries Associated with Dogs and Cats – United States, 2001–2006', Centers for Disease Control and Prevention, 27 March 2009, http://www.cdc.gov/mmwr/preview/mmwrhtml/mm 5811a1.htm. On a lighter note, New Zealand dog trainer Mark Vette has taught several ex-rescue dogs to drive, the first being a bearded collie cross called Porter. See Hilary Hanson, 'Dogs Driving Cars: New Zealand SPCA Puts Canines Behind the Wheel', Huffington Post, 6 December 2012, http://www.huffingtonpost.com/2012/12/05/dogs-driving-cars-new-zealand -spca_n_2244476.html.

8. The quote comes from Becky Bailey, 'The Impact of Electronic Media and Communication on Object Relations', Psychoanalysis Online 2: Impact of Technology on Development, Training and Therapy, ed. Jill Savege Scharff (London: Karnac Books, 2015), 15–16.

9. Judith Donath, 'Artificial Pets: Simple Behaviours Elicit Complex Attachments', in The Encyclopedia of Animal Behaviour, ed. Marc Bekoff (Westport, CT: Greenwood Press, 2004), 955–7.

10. For a review, see Deborah Wells, 'The Effects of Animals on Human Health and Well-Being', *Journal of Social Issues* 65 (2009): 523–43.

11. Molly Crossman, Alan Kazdin and Krista Knudson, 'Brief Unstructured Interaction with a Dog Reduces Distress', *Anthrozoös* 26 (2015): 649–59; see also 'Puppy Love to Help Relieve Exam Stress', University of Bristol, 12 May 2015, http://www.bristol.ac.uk/news/2015/may/puppy-room.html.

12. Figure is redrawn from Wolfram Schultz, 'Multiple Reward Signals in the Brain', *Nature Reviews Neuroscience* 1 (2000): 199–207. For abbreviations, see Chapter 7, note 6. 'Goal representation' in the original is shown as 'Acquisition'.

13. Notably, see Henri Julius et al., *Attachment to Pets: An Integrative View of Human–Animal Relationships with Implications for Therapeutic Practice* (Cambridge, MA: Hogrefe, 2013); Meg Daley Olmert, *Made for Each Other: The Biology of the Human–Animal Bond* (Cambridge, MA: Da Capo Press, 2009).

14. For general reviews of the role of oxytocin in human behaviour, see Anne Campbell, 'Oxytocin and Human Social Behaviour', *Personality and Social Psychology Review* 14 (2010): 281–95; Ruth Feldman, 'Oxytocin and Social Affiliation in Humans', *Hormones & Behaviour* 61 (2012): 380–91.

15. The precise route whereby intranasal sprays of oxytocin affect the brain is still not entirely clear. See Simon Evans et al., 'Intranasal Oxytocin Effects on Social Cognition: A Critique', *Brain Research* 1580 (2014): 69–77.

16. Oriana Aragón et al., 'Dimorphous Expressions of Positive Emotion: Displays of Both Care and Aggression in Response to Cute Stimuli', *Psychological Science* 26 (2015): 259–73; Gideon Nave, Colin Camerer, and Michael McCullough, 'Does Oxytocin Increase Trust in Humans? A Critical Review of Research', *Perspectives on Psychological Science* 10 (2015): 772–89; R. Becket Ebitz and Michael Platt, 'An Evolutionary Perspective on the Behavioural Consequences of Exogenous Oxytocin Application', *Behavioural Neuroscience* 7 (2014), article 225.

17. Evan Maclean and Brian Hare, 'Dogs Hijack the Human Bonding Pathway: Oxytocin Facilitates Social Connections Between Humans and Dogs', *Science* (Washington) 348 (2015): 280–81. For a contrasting view, see Clive Wynne, 'Did Dogs Hack the Oxytocin Love Circuit: Comparing Dogs to Wolves Doesn't Necessarily Inform About Domestication', *Psychology Today*, April 2015, https://www.psychologytoday.com/blog/dogs-and-their-people/201504/did-dogs-hack-the-oxytocin-love-circuit.

18. Johannes Odendaal and Roy Meintjes, 'Neurophysiological Correlates of Affiliative Behaviour Between Humans and Dogs', *Veterinary Journal* 165 (2003): 296–301; Suzanne Miller et al., 'An Examination of Changes in Oxytocin Levels in Men and Women Before and After Interaction with a Bonded Dog', *Anthrozoös* 22 (2009): 31–42.

19. For a review, see India Morrison, Line Löken and Håkan Olausson, 'The Skin as a Social Organ', *Experimental Brain Research* 204 (2010): 305–14; see also Femke Van Horen and Thomas Mussweiler, 'Soft Assurance: Coping with Uncertainty Through Haptic Sensations', *Journal of Experimental Social Psychology* 54 (2014): 73–80; Steve Guest et al., 'Sensory and Affective Judgments of Skin During Inter- and Intrapersonal Touch', *Acta Psychologica* 130 (2009): 115–26.

20. Kenneth Tai, Xue Zheng and Jayanth Narayanan, 'Touching a Teddy Bear Mitigates Negative Effects of Social Exclusion to Increase Prosocial Behaviour', *Social Psychological and Personality Science* 2 (2011): 618–26.

21. Dan-Mikael Ellingsen et al., 'The Neurobiology Shaping Affective Touch: Expectation, Motivation, and Meaning in the Multisensory Context', *Frontiers in Psychology* 6 (2016), article 1986.

22. For further information, see James Curley and Eric Keverne, 'Genes, Brains and Mammalian Social Bonds', *Trends in Ecology and Evolution* 20 (2005): 561–7; see also Lauren Brent et al., 'The Neuroethology of Friendship', *Annals of the New York Academy of Sciences* 1316 (2014): 1–17.

23. Anna Machin and Robin Dunbar, 'The Brain Opioid Theory of Social Attachment: A Review of the Evidence', *Behaviour* 148 (2011): 985–1025; Alessandro Colasanti et al., 'Opioids and Anxiety', *Journal of Psychopharmacology* 25 (2011): 1415–33.

24. See 'Freestyle Dancing with Dogs at Crufts', video posted to YouTube by the *Telegraph*, 29 October 2007, https://www.youtube.com/watch?v=RW6j-K5v_jw.

25. Michael Miller and William Fry, 'The Effect of Mirthful Laughter on the Human Cardiovascular System', *Medical Hypotheses* 73 (2009): 636–9.

26. The quote is from Robert Mitchell and Kristi Sinkhorn, 'Why Do People Laugh During Dog–Human Play Interactions?', *Anthrozoös* 27 (2014): 235–50; see also Robin Maria Valeri, 'Tails of Laughter: A Pilot Study Examining the Relationship Between Companion Animal Guardianship (Pet Ownership) and Laughter', *Society & Animals* 14 (2006): 275–93.

27. For the beta endorphin data, see Odendaal and Meintjes, 'Neurophysiological Correlates of Affiliative Behaviour Between Humans and Dogs'.

28. Similar changes in oxytocin occur in dogs when they're interacting with their owners. In one study, eighteen owners stroked and talked to their dogs for about a quarter hour. At the end of the session, the oxytocin in the dogs'

blood had increased fivefold. Even being stroked by someone familiar to the dog (but not the owner) can cause a similar increase. A single spray of oxytocin into a female dog's nose increases the amount of time she spends looking at her owner. A slight variation in the oxytocin receptor seems to affect the friendliness of both dogs and cats. The similarity between these effects and those that occur in the same situations in owners is, of course, a coincidence, based on a shared mammalian physiology. Neither party can be aware of what hormones are changing in the other – only the senses can transmit feedback. See Odendaal and Meintjes, 'Neurophysiological Correlates of Affiliative Behaviour Between Humans and Dogs'; see also Therese Rehn et al., 'Dogs' Endocrine and Behavioural Responses at Reunion Are Affected by How the Human Initiates Contact', *Physiology & Behaviour* 124 (2014): 45–53; Minori Arahori et al., 'The Oxytocin Receptor Gene (OXTR) Polymorphism in Cats (*Felis catus*) Is Associated with "Roughness" Assessed by Owners', *Journal of Veterinary Behaviour* 11 (2016): 109–12.

Chapter 9

1. For the effects of fashion on the popularity of breeds, see Stefano Ghirlanda et al., 'Fashion vs. Function in Cultural Evolution: The Case of Dog Breed Popularity', *PLoS ONE* 8, no. 9 (2013): e74770, doi: 10.1371/journal .pone.0074770.

2. Linguist Professor Steven Pinker has provided a robust defence of evolutionary psychology and a rebuttal of *tabula rasa*: Steven Pinker, 'Human Nature and the Blank Slate', TED, February 2003, http://www.ted.com/talks /steven_pinker_chalks_it_up_to_the_blank_slate.

3. See Chapter 3 in Robin Dunbar, *Human Evolution* (London: Pelican Books, 2014).

4. Kate Teffer and Katerina Semendeferi, 'Human Prefrontal Cortex: Evolution, Development, and Pathology', in *Evolution of the Primate Brain*, ed. M. A. Hofman and D. Falk, 191–218. *Progress in Brain Research* 195 (2012). For accounts of how evolutionary concepts have been used and misused in psychology, see David Barash, *Homo Mysterious: Evolutionary Puzzles of Human Nature* (New York: Oxford University Press, 2012); Kevin Laland and Gillian Brown, *Sense and Nonsense: Evolutionary Perspectives on Human Behaviour*, 2nd edn (Oxford: Oxford University Press, 2011).

5. Stephen Kellert and Edward O. Wilson, *The Biophilia Hypothesis* (Washington, DC: Island Press, 1995), 416.

6. Roger Ulrich, *Biophilia, Biophobia and Natural Landscapes*, Chapter 3 in Kellert and Wilson, *The Biophilia Hypothesis*.

7. Ke-Tsung Han, 'Responses to Six Major Terrestrial Biomes in Terms of Scenic Beauty, Preference, and Restorativeness', *Environment and Behaviour* 39 (2007): 529–56; Mathew White et al., 'Do Preferences for

Waterscapes Persist in Inclement Weather and Extend to Sub-aquatic Scenes?', *Landscape Research* 39 (2014): 339–58.

8. 'Evidence Statement on the Links Between Natural Environments and Human Health', Department of Environment, Food and Rural Affairs, September 2016, https://beyondgreenspace.net/2017/03/09/defra -evidence-statement-on-the-links-between-natural-environments-and -human-health. For the Toronto study, see Omid Kardan et al., 'Neighborhood Greenspace and Health in a Large Urban Center', *Scientific Reports* 5 (2015), article 11610.

9. For reviews of the effects of natural landscapes on mental well-being, see Jules Pretty, 'How Nature Contributes to Mental and Physical Health', *Spirituality and Health International* 5 (2004): 68–78; Lucy Keniger et al., 'What Are the Benefits of Interacting with Nature?', *International Journal of Environmental Research and Public Health* 10 (2013): 913–35; Danielle Shanahan et al., 'The Health Benefits of Urban Nature: How Much Do We Need?', *BioScience* 65 (2015): 476–85.

10. Richard Fuller et al., 'Psychological Benefits of Greenspace Increase with Biodiversity', *Biology Letters* 3 (2007): 390–94.

11. Arialdi Miniño et al., 'Deaths: Injuries, 2002', *National Vital Statistics Reports* 54, no. 10 (2006).

12. Lynne Isbell, 'Snakes as Agents of Evolutionary Change in Primate Brains', *Journal of Human Evolution* 51 (2006): 1–35; Vanessa LoBue, David Rakison, and Judy DeLoache, 'Threat Perception Across the Life Span: Evidence for Multiple Converging Pathways', *Current Directions in Psychological Science* 19 (2010): 375–9; Joshua New and Tamsin German, 'Spiders at the Cocktail Party: An Ancestral Threat That Surmounts Inattentional Blindness', *Evolution & Human Behaviour* 36 (2015): 165–73.

13. The quotation comes from Ryan DeMares, 'Human Peak Experience Triggered by Encounters with Cetaceans', *Anthrozoös* 20 (2000): 89–103.

14. The quotation comes from André Fiedeldey, 'Wild Animals in a Wilderness Setting: An Ecosystemic Experience?', *Anthrozoös* 7 (1994): 113–23.

15. Desmond Morris, 'Must We Have Zoos?', *Life*, 8 November 1968, 78; Olin Myers Jr, Carol Saunders and Andrej Birjulin, 'Emotional Dimensions of Watching Zoo Animals: An Experience Sampling Study Building on Insights from Psychology', *Curator* 47 (2004): 299–321.

16. Steven Pinker, *How the Mind Works* (New York: W. W. Norton & Co., 1997), 30–31.

17. Douglas Medin and Scott Atran, 'The Native Mind: Biological Categorization and Reasoning in Development and Across Cultures', *Psychological Review* 111 (2004): 960–83.

18. Charles Darwin, *On the Origins of Species by Means of Natural Selection* (London: Murray, 1859), 431.

19. For two views of how 'cell memory' might affect transplant recipients, see Lizette Borreli, 'Can An Organ Transplant Change a Recipient's Personality? Cell Memory Theory Affirms "Yes"', *Medical Daily*, 9 July 2013, http://www.medicaldaily.com/can-organ-transplant-change-recipients-personality-cell-memory-theory-affirms-yes-247498; Joe Shute, 'The Life-Saving Operations That Change Personalities', *Telegraph*, 6 February 2015, http://www.telegraph.co.uk/news/health/news/11393771/The-life-saving-operations-that-change-personalities.html.

20. J. Kiley Hamlin, Karen Wynn and Paul Bloom, 'Social Evaluation by Preverbal Infants', *Nature* (London) 450 (2007): 557–60; Peipei Setoha et al., 'Young Infants Have Biological Expectations About Animals', *Proceedings of the National Academy of Sciences of the United States of America* 110 (2013): 15937–42.

21. Florian Mormann et al., 'A Category-Specific Response to Animals in the Right Human Amygdala', in *Nature Neuroscience* 14 (2011): 1247–9, doi: 10.1038/nn.2899, plus supplemental data at *Nature Neuroscience*.

22. Alex Clarke and Lorraine Tyler, 'Understanding What We See: How We Derive Meaning from Vision', *Trends in Cognitive Sciences* 19 (2015): 677–87; Marieke Mur, 'What's the Difference Between a Tiger and a Cat? From Visual Object to Semantic Concept via the Perirhinal Cortex', *Journal of Neuroscience* 34 (2014): 10462–4.

23. Joshua New, Leda Cosmides and John Tooby, 'Category-Specific Attention for Animals Reflects Ancestral Priorities, Not Expertise', *Proceedings of the National Academy of Sciences of the United States of America* 104 (2007): 16598–603.

24. Hussein Isack and Heinz-Ulrich Reyer, 'Honeyguides and Honey Gatherers: Interspecific Communication in a Symbiotic Relationship', *Science* (Washington) 243 (1989): 1343–6.

25. I take much of my account of the evolution of the human brain from Professor Steven Mithen, *The Prehistory of the Mind: A Search for the Origins of Art, Religion and Science* (London: Thames & Hudson, 1996).

Chapter 10

1. For further details of the Ojibwa and Nayaka, see Nurit Bird-David, '"Animism" Revisited: Personhood, Environment, and Relational Epistemology', *Current Anthropology* 40 (1999): S67–91.

2. Notwithstanding occasional fanciful claims that animals have 'domesticated' other animals. For an example, see Sarah Griffiths, 'Are Monkeys Domesticating WOLVES? Unlikely Relationship May Echo How Dogs Were First Tamed by Humans', *Daily Mail Online*, 15 June 2015, http://www.dailymail.co.uk/sciencetech/article-3115177/Are-monkeys-domesticating-WOLVES-Unlikely-relationship-echo-dogs-tamed

-humans.html. For the changes in thinking possibly required for domestication, see Jean-Denis Vigne, 'The Origins of Animal Domestication and Husbandry: A Major Change in the History of Humanity and the Biosphere [Les Origines de la domestication des animaux et de l'élevage: Un changement majeur dans l'histoire de l'humanité et de la biosphere]', *Comptes Rendus Biologies* 334 (2011): 171–81.

3. Francis Galibert et al., 'Towards Understanding Dog Evolutionary and Domestication History [Histoire de la domestication du chien]', *Comptes Rendus Biologies* 334 (2011): 190–96.

4. Geologically speaking, the Holocene is a warm interglacial period within the ongoing Pliocene–Quaternary glaciations.

5. Science's conception of how domestication worked has changed dramatically over the past decade. See Carlos Driscoll, David Macdonald and Stephen J. O'Brien, 'From Wild Animals to Domestic Pets, an Evolutionary View of Domestication', *Proceedings of the National Academy of Sciences of the United States of America* 106 (2009): 9971–8; Vigne, 'The Origins of Animal Domestication and Husbandry'.

6. Greger Larson et al., 'Worldwide Phylogeography of Wild Boar Reveals Multiple Centers of Pig Domestication', *Science* (Washington) 307 (2005): 1618–21; Laurent Frantz et al., 'Evidence of Long-Term Gene Flow and Selection During Domestication from Analyses of Eurasian Wild and Domestic Pig Genomes', *Nature Genetics* 47 (2015): 1141–9.

7. Interbreeding with wild ancestors would not have been deleterious for all species, one example being horses and other animals used for transport. See Fiona Marshall et al., 'Evaluating the Roles of Directed Breeding and Gene Flow in Animal Domestication', *Proceedings of the National Academy of Sciences of the United States of America* 111 (2014): 6153–8.

8. For a summary of the factors contributing to the scarcity of protein in early agricultural societies, see Jared Diamond, 'Agriculture: The Worst Mistake in the History of the Human Race', *Discover*, May 1987, http://discovermagazine.com/1987/may/02-the-worst-mistake-in-the -history-of-the-human-race. For a discussion of Diamond's article, see Jason Antrosio, 'Agriculture as "Worst Mistake in the History of the Human Race"?', *Living Anthropologically*, 3 February 2013, http://www .livinganthropologically.com/anthropology/agriculture-as-worst-mistake-in -the-history-of-the-human-race.

9. Todd Bersaglieri et al., 'Genetic Signatures of Strong Recent Positive Selection at the Lactase Gene', *American Journal of Human Genetics* 74 (2004): 1111–20.

10. For further discussion of these alternative explanations for domestication of herding animals, see Vigne, 'The Origins of Animal Domestication and Husbandry'; Jean-Denis Vigne and Daniel Helmer, 'Was Milk a "Secondary Product" in the Old World Neolithisation Process? Its Role in the

Domestication of Cattle, Sheep and Goats', *Anthropozoologica* 42 (2007): 9–40.

11. For an account of the revision of the cat's domestication, see Carlos Driscoll et al., 'The Taming of the Cat', *Scientific American* 300, no. 6 (June 2009): 68–75.

12. For a review, see Robert Wayne and Bridgett vonHoldt, 'Evolutionary Genomics of Dog Domestication', *Mammalian Genome* 23 (2012): 3–18.

13. See ibid.'

Chapter 11

1. Peter P. Marra and Chris Santella, *Cat Wars: The Devastating Consequences of a Cuddly Killer* (Princeton, NJ: Princeton University Press, 2016); see also Erik Assadourian, 'Are Pets Bad for the Environment?', *Guardian*, 1 May 2014, https://www.theguardian.com/sustainable-business/reduce-pets-sustainable-future-cats-dogs. For moral stances on the impacts of pet keeping, see Chapter 14 of Peter Sandøe, Sandra Corr and Clare Palmer, *Companion Animal Ethics*. UFAW Animal Welfare Series (Oxford: Wiley, 2016).

2. Relampago Furioso, 'Animals Now Used as Stand-In Children', *The New Modern Man*, 1 February 2016, https://relampagofurioso.com/2016/02/01/animals-now-used-as-stand-in-children.

3. For the classic paper on brood parasitism by the European cuckoo, see Nick Davies and Michael Brooke, 'Cuckoos Versus Reed Warblers: Adaptations and Counteradaptations', *Animal Behaviour* 36 (1988): 262–84; for a recent open-access update, see Rose Thorogood and Nick Davies, 'Combining Personal with Social Information Facilitates Host Defences and Explains Why Cuckoos Should Be Secretive', *Scientific Reports* 6 (2016), article 19872, doi: 10.1038/srep19872. David Noble et al., 'The Red Gape of the Nestling Cuckoo (*Cuculus canorus*) Is Not a Supernormal Stimulus for Three Common Hosts', *Behaviour* 136 (1999): 759–77, first reported the research on 'gape' colour.

4. John Archer has most cogently put forward the 'pets as parasites' perspective, for example in 'Pet Keeping: A Case Study in Maladaptive Behaviour', in *The Oxford Handbook of Evolutionary Family Psychology*, ed. Catherine Salmon and Todd Shackleford, 281–96 (New York: Oxford University Press, 2011); see also John Archer, 'Why Do People Love Their Pets?', *Evolution & Human Behaviour* 18 (1997): 237–57. The editors of the *Oxford Handbook* balanced Archer's account by including a second chapter on pet keeping by James Serpell and Elizabeth Paul, 'Pets in the Family: An Evolutionary Perspective' (297–309).

5. Science fiction's interest in extrasensory perception and telepathy peaked in the 1950s: James Wallace Harris, 'Has Telepathy Become an

Extinct Idea in Science Fiction?', *Auxiliary Memory*, 17 February 2015, http://auxiliarymemory.com/2015/02/17/has-telepathy-become-an-extinct-idea-in-science-fiction.

6. For a summary of the Gallup poll, see 'Many Pet Owners' Paws on Million-Dollar Matter', *Dayton Business Journal*, 28 January 2002, http://www.bizjournals.com/dayton/stories/2002/01/28/tidbits.html.

7. Brooke Scelza and Joan Silk, 'Fosterage as a System of Dispersed Cooperative Breeding: Evidence from the Himba', *Human Nature* 25 (2014): 448–64; Joan Silk, 'Human Adoption in Evolutionary Perspective', *Human Nature* 1 (1990): 25–52.

8. On Doug the Pug, see Brian Koerber, 'Internet Star Doug the Pug Lands a Book Deal', Mashable, 8 February 2016, http://mashable.com/2016/02/08/doug-the-pug-book.

9. For the different motivations that get cats and dogs playing with toys, see my *Cat Sense* and *Dog Sense* (New York: Basic Books, 2013 and 2011, respectively).

10. See William Helton, ed., *Canine Ergonomics: The Science of Working Dogs* (Boca Raton, FL: CRC Press, 2009). For the portrayal of dogs in religion, see Sophia Menache, 'Dogs: God's Worst Enemies?', *Society & Animals* 5 (1997): 23–44.

11. The evolutionary pressures on modern Western populations are highly complex and difficult to interpret. For example, some studies have indicated that (mainly due to differential birth rates) some populations appear to be moving towards *less* education, *lower* income in women, and *decreased* intelligence, quite the opposite of what the 'march of civilization' would predict. See Stephen Stearns et al., 'Measuring Selection in Contemporary Human Populations', *Nature Reviews Genetics* 11 (2010): 611–22.

12. For the 'pets as practise for motherhood' idea, see James Serpell and Elizabeth Paul, 'Pets in the Family, (note 4).

13. For the idea that parental choice drove sexual selection in humans, see Menelaos Apostolou, 'Sexual Selection Under Parental Choice: The Role of Parents in the Evolution of Human Mating', *Evolution & Human Behaviour* 28 (2007): 403–9.

14. See ibid.; Sigal Tifferet et al., 'Dog Ownership Increases Attractiveness and Attenuates Perceptions of Short-Term Mating Strategy in Cad-like Men', *Journal of Evolutionary Psychology* 11 (2013): 121–9.

15. Gene-culture co-evolution has probably done more to shape the human mind than any other adaptive process, and cultural influences on sexual selection may be a particularly powerful subset of this. See, for example, Geoffrey Miller, 'How Mate Choice Shaped Human Nature: A Review of Sexual Selection and Human Evolution', in *The Handbook of Evolutionary Psychology: Ideas, Issues and Applications*, ed. Charles Crawford and Dennis

Krebs, 87–130 (Mahwah, NJ: Lawrence Erlbaum Associates, 1998); Kevin Laland, 'Sexual Selection with a Culturally Transmitted Mating Preference', *Theoretical Population Biology* 45 (1994): 1–15; Ruth Mace, 'The Evolutionary Ecology of Human Family Size', in *The Oxford Handbook of Evolutionary Psychology*, ed. Robin Dunbar and Louise Barrett, 383–96 (Oxford: Oxford University Press, 2007).

16. See 'Scottish Wildcat Conservation Action Plan', Scottish Natural Heritage, 2013, http://www.snh.org.uk/pdfs/publications/wildlife/wildcat conservationactionplan.pdf; 'Protected Mammals – Scottish Wildcat', Scottish Natural Heritage, 22 August 2016, http://www.snh.gov.uk/protecting -scotlands-nature/protected-species/which-and-how/mammals/wildcat -protection. For wolf hybrids, see Martha Schindler Connors, 'Do Wolf-dogs Make Good Pets?', *The Bark*, http://thebark.com/content/do-wolfdogs -make-good-pets, which concludes with the following editors' note: 'In our opinion, despite their undeniable beauty and appeal, deliberately breeding or purchasing wolfdogs as companion animals does a disservice to both *Canis lupus* and *Canis lupus familiaris* as well as to the individual animal. If you love wolves, honour their ancient connection with our domestic dogs by joining the effort to preserve their habitat and maintain their status as a federally protected species.'

17. See Chapter 16 of Frederick Zeuner, *A History of Domesticated Animals* (New York: Harper & Row, 1963). There were also other, cultural barriers to the spread of particular domesticated species, as exemplified by the ancient Japanese nation's embrace of domestic cats but refusal to admit dogs.

Conclusion

1. Melissa Hogenboom, 'Neanderthals Could Speak like Modern Humans, Study Suggests', *BBC News*, 20 December 2013, http://www.bbc .co.uk/news/science-environment-25465102.

2. See Chapter 4 of Robin Dunbar, *Human Evolution* (London: Pelican 2014); Nina Jablonski, 'The Naked Truth', *Scientific American* 302, no. 2 (February 2010): 42–9; Tamás Dávid-Barrett and Robin Dunbar, 'Bipedality and Hair Loss in Human Evolution Revisited: The Impact of Altitude and Activity Scheduling', *Journal of Human Evolution* 94 (2016): 72–82.

3. For an overview of differences between men and women in how they say they feel about animals, see Hal Herzog, 'Gender Differences in Human–Animal Interactions: A Review', *Anthrozoös* 20 (2007): 7–21.

4. Such genes may have existed in our hominid forebears and become more common as they were selected for in *Homo sapiens*. However, it is also possible that some appeared via duplication and/or mutation of existing genes or even via originally non-coding areas of DNA. One estimate has

twenty brain-active genes appearing *de novo* since our split from the chimpanzee lineage; see Dong-Dong Wu, David Irwin and Ya-Ping Zhang, 'De Novo Origin of Human Protein-Coding Genes', *PLoS Genetics* 7 (2011): e1002379, doi: 10.1371/journal.pgen.1002379.

5. @cox_tom on Twitter, 13 March 2016.

6. Michael Balter, 'Are Human Brains Still Evolving? Brain Genes Show Signs of Selection', *Science* (Washington) 208 (2005): 1662–3.

7. For the heritability of personality and other stable differences between individual people, see Eric Turkeimer, 'Three Laws of Behaviour Genetics and What They Mean', *Current Directions in Psychological Science* 9 (2000): 160–64. *First law*: All human behavioural traits are heritable. *Second law*: The effect of being raised in the same family is smaller than the effect of genes. *Third law*: A substantial portion of the variation in complex human behavioural traits is not accounted for by the effects of genes or families. For an update, see Robert Plomin et al., 'Top 10 Replicated Findings from Behavioural Genetics', *Perspectives on Psychological Science* 11 (2016): 3–23; see also Judith Rich Harris, *No Two Alike: Human Nature and Human Individuality* (New York: W. W. Norton & Co., 2006).

8. For the contrary view that pet keeping is mainly a cultural artifact, see Hal Herzog, 'Biology, Culture, and the Origins of Pet-Keeping', *Animal Behaviour & Cognition* 1 (2014): 296–308.

9. For the recent study, see Mark Wade, Thomas Hoffmann, and Jennifer Jenkins, 'Gene–Environment Interaction Between the Oxytocin Receptor (OXTR) Gene and Parenting Behaviour on Children's Theory of Mind', *Social Cognitive and Affective Neuroscience* 10 (2015): 1749–57. For an overview, see Lars Penke and Markus Jokela, 'The Evolutionary Genetics of Personality Revisited', *Current Opinion in Psychology* 7 (2016): 104–9.

10. Claudia Hammond, 'How Being Alone May Be the Key to Rest', *BBC News Magazine*, 27 September 2016, http://www.bbc.co.uk/news/magazine-37444982.

11. For further discussion, see my 'Pets as Ambassadors?', *Psychology Today*, 13 August 2015, https://www.psychologytoday.com/blog/pets-and-their-people/201508/pets-ambassadors. For the original research, see Béatrice Auger and Catherine Amiot, 'Testing and Extending the Pets as Ambassadors Hypothesis: The Role of Contact with Pets and Recategorization Processes in Predicting Positive Attitudes Towards Animals', *Human–Animal Interaction Bulletin* 5 (2017): 1–25.

Index